最想學會的外國菜

全世界美食一次學透透 （中英對照）

作者	洪白陽
攝影	廖家威
翻譯	施如瑛
執行編輯	易師恩、彭文怡、盧蓮芝
校對	馬格麗、易師恩
美術編輯	鄭雅惠
企畫統籌	李 橘
總編輯	莫少閒
出版者	朱雀文化事業有限公司
地址	北市基隆路二段13-1號3樓
電話	（02）2345-3868
傳真	（02）2345-3828
劃撥帳號	19234566 朱雀文化事業有限公司
e-mail	redbook@ms26.hinet.net
網址	http://redbook.com.tw
總經銷	大和書報圖書股份有限公司（02）8990-2588
ISBN	978-986-6780-80-6
CIP	427.1
初版一刷	2010.11
初版四刷	2015.09
定價	350元
出版登記	北市業字第1403號

About買書：

●朱雀文化圖書在北中南各書店及誠品、金石堂、何嘉仁等連鎖書店，以及博客來、讀冊、PC HOME等網路書店均有販售，如欲購買本公司圖書，建議你直接詢問書店店員，或上網採購。如果書店已售完，請電洽本公司。

●● 至朱雀文化網站購書（http://redbook.com.tw），可享85折起優惠。

●●● 至郵局劃撥（戶名：朱雀文化事業有限公司，帳號19234566），掛號寄書不加郵資，4本以下無折扣，5～9本95折，10本以上9折優惠。

Exotic Cuisines

最想學會的外國菜

全世界美食一次學透透 (中英對照)

The highest desire you want to learn.

作者 洪白陽（CC老師）

r

朱雀文化

好吃好吃，就是這麼好吃！

品味各式美食是CC人生最大的樂趣，如果每個人一生都有一項功課要完成，
這個讓人又愛又恨的玩意兒，應該就是我今生的課題吧！
我愛任何讓人帶來幸福味的美食，一碰觸到它，
彷彿人世間再也沒有什麼不如意的事了。
我恨這種種讓人欲罷不能的美食，一碰觸到它，大快朵頤之餘，
體重只會節節高升。唉！CC這種身材再這樣給他吃下去那怎麼得了呀！
但美食讓CC結交到很多好朋友，也因為美食給了CC很多理由到國內外吃喝玩耍，
更因為以美食為職志、教學授課，
使得CC對各類美食都能涉獵與深入了解，且不斷練就更好的廚藝。

多年的授課生涯，讓CC了解到我的學生、也就是一般愛做菜的大多數人，
喜愛的料理方式，無非就是「簡單＋美味」的料理。
如何化繁為簡、將口味調整為讓大家都能接受的風格，就是這本書出版的目的了。
希望藉由這本書，讓讀者能認識世界各地的異國料理，多多嘗試各種辛香料
組合出來的誘人風味，帶給家人朋友耳目一新的菜單吧！

這本書的編輯問我：怎麼書裡就一直出現「好吃」、「好吃」。
我跟他們說：就真的很好吃呀！果然在拍攝期間，每拍完一道美食，
編輯和攝影師拿起湯匙舀菜入口後，就是這一句：「好吃」。
我愛這本食譜，將這本食譜獻給愛我料理的人！

To try out a wide variety of cuisines is CC's greatest joy in life. If everyone has their own homework to finish, this thing that drives me both mad and happy should be my homework for this lifetime.

I love any cuisine that brings people happiness. Once you come into contact with such a cuisine, there seems to be no more misery in the world

I hate the enormous variety of cuisines. Once you encounter it, you only increase in size. Ah, what is going to happen to a figure like CC's?

However, through delicious cuisine CC has made so many good friends. Because of it, CC has had lots of excuses to go everywhere to enjoy delicious food both at home and abroad. Further, since delicious cuisine has become her future career aspiration and teaching goal, CC has enjoyed the chance to delve further into cooking and study all kinds of delicious food. Since then she has sharpened her cooking skills.

Many years of teaching has enabled CC to realize that her students, which also means most people who love cooking, prefer cooking methods to be simple and delicious. Making complicated process simple and adjusting the flavor to the style that people can accept are the purposes of publishing this book. Hopefully readers can become familiar with authentic dishes from around the world through this book, and try out the charming aroma of a variety of spices and herbs working together. Let's create a totally new menu for your family and friends!

The editor of this book asked me, why the word "delicious" keeps appearing in the book. I told them it is the truth! During the photo shoot of the dishes, every time a dish was finished, the editor and photographer would grab their spoons and taste, and the word that came out of their mouth is "delicious".

I love this cookbook, and dedicate this book to the people who love my cooking!

特別感謝助手：吳杰修

全世界美食一次學透透
最想學會的外國菜
Contents

最想學會的外國菜
Part1 蔬果、前菜類
Appetizer Section

泰式月亮蝦餅

全世界美食一次學透透
最想學會的外國菜

繽紛烤全雞

里昂燴鮮蝦

最 想 學 會 的 外 國 菜
Part4 **主食和飯類**
Main Dish and Rice Sections

最 想 學 會 的 外 國 菜
Part5 **湯品**
Soup Section

黃金鮮蝦飯

馬賽海鮮湯

這些小材料就是讓食物更香更好吃的魔法師

要讓料理好吃，手藝火候最重要，
好的食物和輔助調味料更是成功的保證。以下這些小材料就是CC美食廚房裡的常備品，
尤其要做出美味的異國料理，辛香料和起司更是不可缺少的關鍵食材；一起來認識吧～

These ingredients are the magic ingredients that make the dishes even more aromatic and delicious
The skill of cooking and the control of heat are very essential to a successful dish. Of course, good ingredients and seasonings help a lot too. The following ingredients are the necessities found in CC's kitchen. Spices, herbs and cheeses are the key to delicious authentic dishes, Let's get to know them~~

匈牙利紅椒粉

適合用於燉湯或烤肉等，或想讓菜餚上色但不吃辣時。還有一個功能，就是家中沒有烤箱，但想在食物上呈現出有點焦黃的感覺，可以撒些匈牙利紅椒粉，會有微焦的感覺喔！

Hungarian Paprika: Suitable in stew broths or barbecuing pork, or making the dish more colorful without adding spiciness. Another function: if an oven is not available but you want to present the food with a yellowed, baked appearance, you can sprinkle with a little paprika.

鬱金香粉

也就是薑黃粉，是調製咖哩粉成分中最重要的一個。常用於印度菜或東南亞料理。

Turmeric powder: one of the most important ingredients in preparing curry powder. It is often used in Indian dishes or South East Asian dishes.

蕃茄糊

常用在煮湯或燉肉，最常用於義大利肉醬或披薩餅皮上。

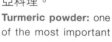

Tomato paste: it is often used in stew broths or meat, especially in an Italian-style meat sauces for pasta or on top of pizza dough.

泰國打拋醬

內含九層塔、辣椒、蒜頭等辛香料組合成的醬料，常用於炒雞肉、豬肉、牛肉或沾醬。也可以醃製烤雞，炒飯、拌麵都適合。

Thai Dapao Paste: a combination of basil, chili pepper, garlic, and other spicy herbs. It is used in sautéed chicken, pork, beef or dipping sauce. It can be a marinade for roasted chicken, or fried rice, or mixed with noodles.

甜菜罐頭/甜菜

著名的羅宋湯就是以它為主要的顏色和味道，適合用來煮湯。台灣目前也買得到新鮮的甜菜根了，而罐頭製品是保存及處理都比較方便。

Canned Beets/Beets: The famous Taiwanese style beef-flavored pork soup (luo song tang) uses it as main color and flavor. It is perfect for soups. Fresh beets are available in Taiwan, but canned beets are more convenient for storing and handling.

牛肝菌

常用於歐式料理的菇類，乾燥的香氣較濃郁。使用範圍極廣，可運用在燉肉、燉飯、煮湯，和拌炒義大利麵。也可做為牛排等肉類的配菜和醬汁。

Boletus edulis: A mushroom used in European dishes. In dried form it has a heavy aroma. Its usage ranges broadly, from stewing meats and rice, to soups, to sautéing pasta. It is also used in side dishes and dipping sauces for meats.

韓國辣椒醬

適合用於調沾醬、燉煮或炒年糕、炒豆腐、烤肉、涼拌菜等；使用範圍很廣。

Korean Chili Paste: a wide variety of uses, such as dipping sauces, stewing, sautéing new year cake, tofu, barbecue meats, and cold vegetable mixes.

泰國辣椒膏

最常用在酸辣蝦湯或酸辣海鮮湯，也常用於泰式涼拌海鮮、炒雞丁碎肉等。CC則喜歡拿來做乾拌麵。

Sriracha: It is often used in sour and spicy shrimp soup, or sour and spicy seafood soup. It is also used in dishes such as cold Thai seafood mix and sautéed chicken. CC loves to have it in mixed dried noodles.

棕櫚糖(椰子糖)

泰國棕櫚糖，又稱香椰糖、椰子糖。泰文為Nam Tan Pip，是從扇椰樹的果實製成乾而成，有種特殊的

清香味，用在各式泰國料理，無論燉、煮、炒、涼拌、甜點都適合。也是東南亞各國常用的食材，咖啡和茶內加入棕櫚糖也很香很好喝喔！

Palm Sugar: Thai palm sugar is also known as coconut sugar, or Nam Tan Pip in Thai. Traditionally it is made from the sap of Phoenix sylvestris, known in English as the Palmyra palm or the date palm. It has a special light fragrance. It is widely used in many Thai dishes and desserts and is a common ingredient in South East Asia. A little palm sugar will make even coffee and tea much better!

韓國泡菜

直接吃或炒肉片、碎肉，涼拌，也可以做成泡菜火鍋，都很適合。

K o r e a n Kimchi: Serve directly or sauté with meat, minced meat, cold mixes, or even in kimchi hot pot.

泰國綠咖哩醬、紅咖哩醬

紅咖哩和綠咖哩基本材料是差不多的，只是紅咖哩加入紅辣椒、綠咖哩加入青辣椒，故顏色

不同，而且綠咖哩較辣、較清爽，紅咖哩則口感比較厚實。綠咖哩常用於燉煮等料理，或快炒式的雞丁海鮮類。紅咖哩適合燉肉或魚鮮料理，一般家庭常用來炒肉末或燒冬瓜、南瓜。也有用於湯料理。

Thai Green Curry, Red Curry: red and green curry have essentially the same basic ingredients, except that red curry uses red chili pepper and green curry uses green chili pepper, which is why the color is different. Green curry is spicier and lighter, while red curry has a thicker texture. Green curry is often used in stews, or in sautéing diced chicken or seafood. Red curry is suitable for stewing meat or cooking fish. Ordinary

families use it to sauté minced meat, or winter melon and pumpkins. Sometimes it is used in soups.

鼠尾草

是去腥味的最佳香料，直接聞氣味不太好聞，但放入料理中卻相當可口，常使用於羊排或魚類料理，也是製作德式香腸重要的香料之一。

Sage: It is the best herb for eliminating "fishy" odors. It does not smell good, yet it is really delicious in dishes. It is often used in lamb or fish dishes, and is a key flavoring in German sausage.

巴薩米克醋

產於義大利的頂級香醋，口感獨特且香氣豐郁，適合用於牛排、羊排、豬排等醬汁，調製沙拉醬、烹炒菇菌類、炒義大利麵也很搭，淋在冰淇淋上口感更是迷人。

Balsamic vinegar: a top class vinegar that produced in Italy. The taste is unique and fragrance is unbelievable. It is good in beef steak, lamb chops, or pork steak sauces. It is perfect for making salad dressings, sautéing mushrooms, or pasta. It is even tasty drizzled over ice cream.

小茴香

最常使用於燉肉料理，有除膩的功能，也是製作滷製食品的必備香料。

Cumin: mostly used in stewed pork dishes, it energizes the body and is a must for preparing stewed dishes.

奧勒岡

常用於披薩皮或燉肉料理。與蕃茄、乳酪搭配絕佳。奧勒岡和羅勒是義大利料理最常使用的兩種香料。

Oregano: it is mostly used in pizza or stewed meat dishes. It matches perfectly with tomato and cheese. Oregano and basil are the most common two herbs in Italian cuisine.

檸檬葉

其實這不是我們常見的檸檬的葉子，而是卡非萊姆（也有稱為亞洲萊姆）的葉子，使用時葉子需撕開才會更香，這是泰式料理不可缺少的香料之一，常用於肉類、海鮮及蔬菜料理。也是製作泰式海鮮湯一定要放的香料。

Kaffir Lime Leaves: ot lemon leaves, but the leaves of Kaffir Lime (also known as Asian Lime). Tear the leaves apart when using them, so that the fragrance will emerge. This is one of the herbs that you need when preparing Thai cuisine. It is often used in meat, seafood and vegetable cuisine, and a crucial herb for Thai seafood soup.

月桂葉

常用在熬製高湯、各種燉肉及燉海鮮上。尤其烹煮牛肉、牛雜能發揮極佳的效果。由於味道稍重，入鍋時不宜放得太多，通常放1片就可以了，否則會蓋住主要食材的原味

Bay Leaf: largely used in stews and soup broths. It is at its best with stewed beef and beef organs. Due to its heavy flavor, do not add too much; usually one leaf is enough, or it will overpower the original flavor of the main ingredients.

丁香

是許多綜合辛香料與咖哩粉的配方之一，適合用於熬製高湯和燉肉。

Cloves: it is one of the ingredients in curry powder, perfect for cooking soup broths or stewing meats.

南薑

是泰式料理最重要的香料之一，相較於我們常使用的薑，南薑較清香且較不辛辣，運用範圍很廣，舉凡煮湯、燉肉、海鮮、涼拌都適合。

Greater Galangal (Alpinia galanga): It is one of the most important spices in Thai dishes. It is a member of ginger family, a rhizome with flavor reminiscent of camphor, light and fragrant, but not as spicy. It features widely in Thai cuisine, especially in soups, stewed meat, seafood or cold mix dishes.

香茅

也是泰國主要的香料之一，使用範圍很廣，煮湯、燉類、涼拌都適合，也有用於飲料上。

Lemon Grass: a common Thai flavoring with a broad range of uses in cooking including soups, stewed dishes, or cold mixes. Sometimes it is used in drinks.

番紅花

全世界最貴的香料之一，適合用於煮湯、燉飯、海鮮醬汁，特別的香味可去腥並有上色的功能，這也是製作西班牙海鮮飯不可或缺的香料。

Saffron: one of the most expensive herbs in the world. It is excellent in soups, stewed rice, and seafood dressings. It has a unique flavor that eliminates unwelcome flavors and colors dishes a lovely golden color. It is a key herb for Spanish seafood rice.

紅胡椒粒

胡椒的一種，漂亮的顏色可增添盤飾的美感，也增加料理中的香味。

Red Peppercorns: a type of peppercorn whose beautiful color can garnish the plate and increase the flavor of the dish.

水參

即新鮮人參，常用於燉湯，在大型百貨公司超市均有賣。

Water Ginseng: Water Ginseng: a kind of fresh ginseng often used in stew broth. It can be purchased at the supermarket in large department stores.

第戎芥末醬

又稱法式芥末醬，和羊肉、牛肉、豬肉的搭配十分契合，在歐式料理常做為醬汁調料的基底，也有加上蜂蜜調成的蜂蜜芥末醬，沾食炸雞最受歡迎。

Dijon Mustard: also known as French mustard. It is a good match with lamb, beef and pork. Dijon mustard often used as a foundation of the dressing in European cuisines. There is also honey mustard, which has honey added, which is the most popular dip for fried chicken.

茵陳蒿

是法式料理常見的香料，味道相當特殊，燉肉或煮湯皆適合，具有消除魚肉類腥味及分解脂肪的作用，新鮮的茵陳蒿則可運用在醬汁或沙拉醬上。很適合將其葉子碎末加在烤魚、烤雞上。此外也可運用在炒蛋、煎蛋等蛋類料理中。

Artemisia: it is often seen in French cooking. It has a very special flavor and is suitable for stewing pork or cooking soup. It removes "fishy" odors and decomposes fat. Fresh Artemisia can be used in sauces or dressings. Its leaves can be chopped and added to roasted fish or chicken. It can also be used in sautéed or fried eggs.

泰國河粉

是泰國美食不可少的主食之一，涼拌、湯河粉、乾拌河粉都很好吃。

Thai pho (rice noodles): one of the main ingredients in Thai cuisine. In cold mixes, pho soup, or dried pho mixes, it is delicious.

酸子醬(羅望子醬)

泰式料理中只要有帶酸的料理幾乎少不了它。如涼拌酸木瓜、涼拌海鮮、酸辣湯等。羅望子長得像龍眼，但酸甜黏稠，是很好吃的水果。

Tamarind Paste: any Thai dishes that are sour, such as cold papaya mix, cold seafood mixes, and sour and spicy soup, are made with tamarind paste. It resembles a longan and tastes sour, sweet and sticky. It is a delicious fruit!

蘋果醇

即蘋果醋，適合製作沙拉醬汁，也可運用於冰砂或冷飲中。

Apple Vinegar: perfect for salad dressings, also use in smoothies and cold drinks.

紅腰豆

原產於南美洲，含豐富鐵質，常吃可補血、增強免疫力，是相當健康的食材；使用於沙拉或燉肉、燉飯皆適合。

Red Kidney Beans: originally from South America. They are rich in iron, and supplement the blood and increase the immune system if eaten often. It is a very healthy ingredient and suitable in salads or stewed meats or rice.

義大利米

最好吃的燉飯就是用這種米，也可運用在湯內或沙拉。

Italian rice: the most delicious rice stew uses this ingredient. It can also be used in soup or salad.

加拿大經典火腿

常運用於沙拉、三明治，也可切小丁撒在湯裡，這是CC覺得製作藍帶豬排時最好吃的食材。

Canadian Classic Ham: Use often in salads and sandwiches. It can also be diced and sprinkled in soups. CC considers it the most delicious ingredient in preparing Blue Belt Pork Steak.

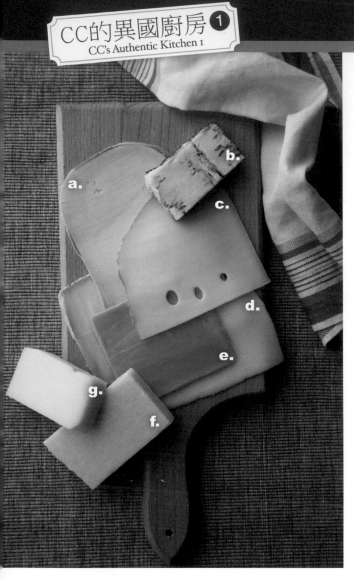

乳酪 Cheese

a. 高達起司

盛產於荷蘭的高達起司屬於半硬質起司，外型以扁圓車輪型聞名，口感較重，外皮包覆有一層薄蠟，而薄蠟的口味不一，其中以棕褐色外皮的煙燻起司，最受歡迎。直接吃或做成沙拉都很適合。

Gouda Cheese: Gouda Cheese, from Holland, is a semi-hardened cheese. It is famous for its flat wheel shape. The flavor is strong and the surface is wrapped with a layer of wax, which can have a variety of flavors. The most popular version of Gouda is the smoked flavor cheese, which has a dark brown color. It is good in salads or eaten directly.

b. 歌魯拱索拉起司

世界三大藍黴乳酪之一，原產地為義大利歌魯拱索拉村，整個村莊以生產乳酪聞名，屬軟質乳酪，熟成期約3個月。適合用於調製醬汁、開胃菜，或與麵包搭配。

Gorgonzola Cheese: One the world's three blue cheeses, it comes from the village of Gorgonzola in Italy. The whole village is famous for producing the cheese. A kind of soft cheese, it takes three months to ripen. It is perfect for for mixing in dressings, appetizers, or serving with breads.

c. 富勒比起司

原產地為法國富勒比，味道溫和清香、質地結實，吃起來頗有嚼勁，直接使用在開胃菜或三明治上，或與紅酒搭著吃，口味絕佳。

Folepi Cheese: Originally from France, it has a mild, light flavor. Its firm texture provides a chewy taste. It can be served as an appetizer, or in a sandwich, or with red wine. The flavor is incredible.

d. 葛瑞爾起司

知名的瑞士乳酪，熟成期約6個月。味道稍濃且帶酸味、帶點臭香，是歷史悠久的乳酪。

Gruyere Cheese: A very famous Swiss cheese that takes six months to ripen. Its flavor is slightly thick and sour. It is sticky yet fragrant. It is a cheese with a long history.

e. 巧達起司

原產於英國，是一種略帶鹹味而香氣濃郁的硬質起司，口感馥郁溫和、帶有淡淡堅果香，金黃色的外表常刨絲入沙拉或料理中，也適合焗烤料理，但巧達起司較沒有牽絲的效果。

Cheddar Cheese: Originally from England, it is a hard cheese that has a little salty flavor, but a strong aroma. The fragrance is mild with light nutty note. This cheese with its pleasing golden color is often shredded in salads or dishes. It is also suitable for baking but has a less silky effect.

f. 帕瑪森起司

屬於硬質乳酪，可說是義大利的國寶乳酪，常磨成粉或是刨成薄片使用，搭配義大利麵食、凱撒沙拉、焗烤，或撒於義大利麵上。

Parmesan Cheese: A hard cheese, it is known as the treasure of Italian cheeses. It is often ground or grated into thin slices to serve on pasta, Caesar salad, and baked dishes.

g. 愛蒙塔起司

口感香醇濃郁，略帶淡淡花香與乾果味。適合運用在沙拉和開胃菜中，或與美酒一起享用；切片夾麵包或放在洋蔥湯上烤都適合。瑞士最有名的起司火鍋就是用愛蒙特和葛瑞爾一起烹煮的。

Emmental Cheese: The cheese is filled with holes, just like the cheese that mice love in cartoons. Its flavor is fragrant and thick, with a light floral scent and dried fruit flavor. It is suitable in salads and appetizers, or served with red wine. It may also be sliced and stuffed in breads or baked atop onion soup.

香草 Herbs

a. 羅勒葉

義大利料理中最佳配角，是製作青醬的主要材料；和乳酪、蕃茄很搭，也常用於海鮮及各式肉類料理。

Basil: The boot supporting character in Italian cuisine. It is the main ingredient in making pesto and goes well with cheese and tomato. It is often used in seafood and all type of meat dishes.

b. 蝦夷蔥

極細的蔥，常用於裝飾，或切碎來做沙拉醬、醬汁、沾醬。

Chives: Very fine chives, used mostly as a garnish, or chopped finely in salad dressings, sauces, or dipping sauces.

c. 巴西里（番茜）

常用於盛盤裝飾，或切碎來做沙拉醬、醬汁、沾醬。

Parsley: used in garnishing plates, or can be chopped finely to make salad dressings, sauces, or dipping sauces.

d. 迷迭香

具有強大的去腥力，大多用於肉類羊肉、魚類料理，尤其是羊排、烤魚。

Rosemary: it has a strong aroma that drowns out unpleasant odors, mostly used in meat dishes, especially lamb and fish dishes such as lamb chops or roasted fish.

e. 薄荷

常用於盛盤裝飾，而在東南亞料理中，與涼拌菜搭配是很爽口的味道。

Mint: often used as a garnish. However, in Southeast Asian cooking, its light and tasty flavor is a good match in cold dishes.

f. 百里香

常與月桂葉一起熬煮高湯，適合運用在各類燉煮料理。可在入鍋前以手拍三下，讓香味溢出再丟入鍋中。

Thyme: often used with bay leaves in making soup broth, it is suitable for any kind of stewed dishes. Press it gently with hands three times to let the flavor released before adding to the pan.

聰明用好鍋和廚房小道具，讓料理事半功倍

CC老師每天在廚房裡煮菜，廚房簡直就是我的第二個臥房；尤其每當炎炎夏日要燉煮熬湯時，可真是揮汗下鋤（廚）呀！所以要以最少的烹煮時間做出最健康美味的料理，就是我研究料理的終極目標之一。

常有人問我為什麼短短幾分鐘就能做出好多道好吃的料理？秘訣就是我的廚房大小法寶。好鍋和好的工具絕對是能讓料理輕鬆的最大武器，我常使用的工具如下，提供讀者參考。

Teacher CC spends most of her time in the kitchen cooking. Her kitchen is like her second bedroom. When the weather is extremely hot she still has to prepare soups, cooking and sweating at the same time. Therefore, using the least amount of time to prepare the most healthy and delicious meals has become an important goal in studying cooking.

Many people have asked me how I can prepare so many delicious dishes in such a short time. The secret lies in my large and small magic weapons in the kitchen. A good pan and superior utensils are the most important weapons in making cooking easier and more relaxing. The utensils I use are described below. Readers can use this as reference !

雙壁鍋

對這個榮獲奧地利發明金牌獎的鍋具CC特別鍾愛，只要食材新鮮，煮什麼都好吃，而且滷豬腳花生可以不必放水。省油、省能源、省時間，這絕對是一只節能減碳的好工具。

Durotherm: This pot won the Austria invention award, which is CC's favorite. Every nature fresh food will be cooked well. and it is not necessary to supply water while you stewed peanut-and-trotter..This pot saves oil, saves energy and saves time. It is a perfect tool that reduces carbon emissions.

壓力鍋

冬天到了難免要燉點補品養身，壓力鍋可以在30分鐘就滴出健康的元氣雞精（以前用電鍋滴，5～6小時），不論是滷牛腱、燉排骨湯、水煮雞都快快快，很適合CC急性的個性。

Duromatic: People tend to make dishes to nurse the body through the cold winter. A healthy, energy-filled chicken essence can be prepared in 30 minutes in a Duromatic pan (it takes about 5to6hours minutes with a rice cooker). Whether stewing beef tendon, pork rib soup, or water-cooked chicken, it is done faster than you can imagine. It is perfect for an impatient person like CC.

休閒鍋

CC幾乎都用這個全能萬用鍋煮飯煮菜，煮飯6分鐘、蒸魚不必放水（只放酒），只要6分鐘，搭配好順序可以在30分鐘裡煮好6道菜，而且全程可以不放油、不洗鍋。放置到保溫外鍋能繼續燜燒，保溫、保冰，是一只全方位料裡好鍋。

Hot Pan: Teacher CC does most of her cooking with this pan. It takes 6 minutes to cook rice, and 6 minutes to steam fish and no need to supply water except wine. and if well planned, 6 dishes can be finished within 30 minutes. Throughout the entire process, it's no need to clean the pan and no need to supply oil.If the pan is dropped onto the insulation pot, it will keep smoldering, insulation and icing. No wonder it is a multi-function cooker (pan).

易拉轉

免插電的食物調理機,輕輕一拉就可以把蔥、薑、辣椒等辛香料攪細,愈拉愈細,切洋蔥從此不流淚,手指也不再殘留蒜味。以此做成的沾醬,香味可提高一倍。

Twist & Chop: This device is a food processor without an electric core. Just pull gently, and ginger, chili pepper, scallions, and other foods, can be finely ground. The more you pull, the finer the ingredients will be. From now on, there will be no more tears from cutting onions, and no more garlic flavor on your finger tips.To use such kind of sauce can increase the aroma higher.

潔能板

放在瓦斯爐上可以使鍋子受熱均勻,烹調食物時,可以節省時間,更使鍋底不用直接接觸火焰,省去刷洗工夫。小鍋加熱、鋼杯熱牛奶、便當盒熱菜、摩卡壺、奶泡壺都可使用,還可解凍。

Energy Saver Protector: Place it on the stove under the pan to make the pan heat more evenly. This saves time when mixing the ingredients of the dish in the pan. The bottom of the pan does not contact the fire directly, meaning that you do not have to clean it. Whether heating up a small pan, warming up milk in a stainless steel cup, heating leftovers in a lunch box, or even a moka pot or a milk foamer, it can be used with anything you can name. And also can unfreeze the food.

快易夾

可當鍋鏟炒菜,可當攪拌打蛋器,可當夾子使用,好握、好洗,不傷鍋。

Easy-Lock Tongs: These can be used as a spatula, egg beater or tongs. They are easy to hold, easy to clean and do not damage the pan.

多用途強力剪刀

設計得相當省力就可以剪食材,可剪雞骨頭,還可以拿來剪花、剪電線,當開瓶器使用,吃螃蟹也OK。

Multipurpose Kitchen Scissors: These are designed to cut ingredients easily. They can cut chicken bones, flowers, or even electric lines, or be used as a can opener. The taste of the sauce match crabs very well.

玉米刮刮樂

整隻玉米造型,一看就知道是用來刮玉米的,可以將玉米粒胚芽部分一次刮乾淨,且刮出來的玉米粒粒飽滿,這個小工具也是很多媽媽和小朋友的最愛。

Corn Zipper: Shaped like an ear of corn, you can tell it is a corn kernel remover right away. It can smoothly remove corn kernels, including the germ from the kernel, leaving full, beautiful corn kernels. This small tool is a favorite of both mothers and children.

蔬果削皮刀

削蕃茄皮、奇異果皮超方便,也適合削其他軟皮水果,如水蜜桃。

Piranha Peelers: This tool is super convenient for peeling tomatoes, kiwis, or other soft-skinned fruits.Like juicy peach.

蔬菜脫水機

做沙拉前菜的最好工具,輕輕一拉即可將蔬菜水分瀝乾。

Salad Spinner: This is the perfect utensil for preparing salads or appetizers. Just give it a gentle whirl and it can thoroughly drain the liquid from vegetables.

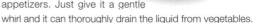

給我一個百花齊放的春天，
我要規畫很多很多去遠方的行程：
去京都看櫻花、到越南買雜貨、
去泰國遊河，還要飛去有陽光的加州、
有藝術氛圍的歐洲……
你問我為什麼要去這裡那裡，不為什麼，
只因為這裡那裡有美食。
好味在哪裡，旅人就在那裡。

What a bloomy spring，
I wanna plan to visit many far-far-away places;
To Kyoto for sakura、To Vietnam for Zakka、
To Tailand for river sightseeing.
I also wanna fly to sunshine California、
to Europe where is filled with art-atmosphere………
If you ask me the reason why I wanna go to such places，
I will tell you: "just no reason!!"
It's only because of the delicacies of these places.
The smart travelers always show up where there are delicacies.

京都風味
百菇齊放沙拉
Kyoto Hundred Mushroom Salad

好吃爽口可提升免疫力，並含有豐富多醣體的清麗爽口沙拉。

This dish is delicious and light, boosts the immune system,
and is also rich in polysaccharides

京都風味百菇齊放沙拉

材料
約5～6種菇類各100g.、蘿蔓生菜200g.、牛蕃茄1個(切塊)

沙拉醬
柴魚醬油5匙、味酥1匙、檸檬汁2匙

做法
1. 將蘿蔓生菜葉片一葉一葉洗淨，入冰水冰鎮20分鐘後取出，以蔬菜脫水機瀝乾水份。
2. 菇類放入鍋內汆燙（使用雙壁鍋不需放水：蓋上鍋蓋以中火煮至冒煙，轉小火續煮約2～3分鐘），取出瀝掉水份。
3. 盤中先鋪上蘿蔓生菜，再放上菇類和牛蕃茄。將沙拉醬拌勻淋上即成。

完美烹調寶典
Tips for a Perfect Dish
1. 蘿蔓生菜洗淨後可放冰箱冰鎮20分鐘，取出瀝乾水分再吃更爽口好吃。
2. 牛蕃茄可以用小蕃茄替代。

Ingredients
100g approximately 5-6 kinds of mushroom, 200g romaine lettuce, 1 beefsteak tomato (cut into pieces)

Dressing
5T bonito soy sauce, 1T mirin, 2T lemon juice

Methods
1. Rinse romaine lettuce leaves one by one, soak in ice water for 20 minutes and remove. Drain with a vegetable dehydrator until dry.
2. Blanch mushrooms in boiling water (In a Durotherm no liquid: Cover and cook over medium low until smoking, reduce heat to low and continue simmering for about 2-3 minutes), remove and drain.
3. Line the serving plate with romaine lettuce leaves, then top with mushrooms and tomato. Drizzle with dressing and serve.

Tips for a Perfect Dish
1. For a crunchy texture: Rinse the iceberg lettuce leaves and soak in ice water in the refrigerator for 20 minutes. Remove and drain well.
2. Beefsteak tomatoes may be used instead of cherry tomatoes.

越式涼拌沙拉
Vietnamese Cold Salad

越南風味的涼拌青木瓜
清爽滋味有別於常見的泰式木瓜沙拉,且充滿著更多層次風味喔!

A Vietnamese style cold green papaya salad whose clear, light taste is
different from Thai style papaya salad, with many layers of flavor!

越式涼拌沙拉

材料

草蝦8隻、雞胸肉200g.、青木瓜200g.(切絲)、碎花生3匙、九層塔適量、薄荷葉10片、紅蔥頭3粒(切絲)

淋醬

白砂糖2匙、魚露2 1/2匙、蒜末1 1/2匙、紅辣椒末1匙、檸檬汁2匙

做法

1. 草蝦去腸泥，放入鍋中氽燙或電鍋蒸熟（使用雙壁鍋不需放水：蓋上鍋蓋以中小火煮至冒煙，轉小火續煮約3分鐘）。待涼後去殼，背部劃開成兩片蝦。

2. 雞胸肉放入電鍋蒸熟或以滾水煮熟（使用雙壁鍋：先加入3匙水，蓋上鍋蓋以中小火煮至冒煙，轉小火續煮約6分鐘），取出待涼，切絲。

3. 製作淋醬：白砂糖加入4匙溫水中拌勻，待涼後再與其他材料拌勻即成。

4. 將青木瓜絲鋪於盤中，鋪上雞絲，上層再放蝦肉，撒上碎花生和紅蔥頭絲，最上面放九層塔和薄荷葉，淋上淋醬即可享用。

完美烹調寶典 *Tips for a Perfect Dish*

1. 這裡的香草都要用新鮮的葉片，沒有薄荷還是很好吃，但九層塔是絕不可缺少的。

2. 淋醬的比例酸的剛剛好，讀者也可以依自己的口味調整檸檬汁的份量。

3. 使用材質好的鍋煮出來的食材較鮮甜，可以吃到食物原本的鮮味。

Ingredients

8 glass shrimp, 200g chicken breast, 200g green papaya (shredded), 3T chopped peanut, basil as needed, 10 mint leaves, 3 shallots (shredded)

Sauce

2T white granulated sugar, 2 1/2T fish sauce, 1 1/2T minced garlic, 1T minced red chili pepper, 2T lemon juice

Methods

1. Blanch shrimp in pan or steam until done in a rice cooker (in a Durotherm no liquid Durotherm: Cover and cook over medium low until smoking, redue heat to low and continue cooking for about 0 minutes). Let rest until cool and remove the shrimp shells, then halve the shrimp along the back.

2. Steam chicken breast in rice cooker until done, or cook in boiling water until done (in a Durotherm: Add 3T of water, reduce heat to low, continue cooking for about 6 minutes). Remove and rest until cool, then shredded.

3. To prepare sauce: Combine white granulated sugar and 4 spoonfuls of warm water well, wait until cool and mix well with the rest of the ingredients.

4. Line shredded papaya at the bottom of the plate, then top with shredded chicken breast, then shrimp meat. Sprinkle with chopped peanut and shallots. Place basil leaves and mint leaves on top. Drizzle with dressing and serve.

Tips for a Perfect Dish

1. Use fresh herb leaves in this recipe. This dish is still good without mint, but basil is a must.

2. The proportion of the sauce is just right. Readers can adjust the amount of lemon juice as desired.

3. The dish is fresher and sweeter with a better quality cooking pan, which enables the real flavor of the food to be tasted.

美味四射涼拌
Delicious Four-Color Salad

這是以滴過雞精的雞肉再利用的美食。
This is the reuse find food of chicken which is ever dropped chicken essence.

材料

雞肉300g.(剝成絲)、小黃瓜2條(切丁)、小蕃茄150g.(切丁)、黃甜椒1個(切丁)、萵苣生菜300g.、嫩薑末11/2匙、蔥末2匙、香菜末3匙、芹菜末4匙

醬料

醬油4匙、蘋果醇2匙、檸檬汁2匙、香油1匙、芝麻醬21/2匙

做法

1. 將嫩薑末、蔥末、香菜末、芹菜末放入調理盆內,與調味料拌勻,待5分鐘後味道會更好。
2. 雞絲、小黃瓜丁、小蕃茄丁、甜椒丁倒入做法1.拌勻。
3. 萵苣生菜洗淨,以蔬菜脫水機瀝乾,鋪盤圍邊,將其他材料擺置盤中,食用時以萵苣生菜包做著吃。

泰式海鮮涼拌
Thai Style Seafood Salad

材 料

透抽1隻、草蝦10隻、淡菜8個、洋蔥1/2個(切絲)、芹菜4支(切段)、紅辣椒2支(切斜段)、蔥2支(切段)、小蕃茄6個(對切)、香菜末2匙

醬 料

泰國辣椒膏2匙、魚露1匙、酸子汁1匙、檸檬汁2 1/2匙、白砂糖2/3匙、蒜末2匙、紅蔥頭3顆(切絲)拌勻

做 法

1. 透抽洗淨去除薄膜切片，草蝦去頭殼留尾巴，由背部劃一刀去除腸泥。

2. 將透抽、草蝦和淡菜放入沸水汆燙熟(使用雙壁鍋不需放水：蓋上鍋蓋以中火煮至冒煙，轉小火續煮約3分鐘)，取出瀝乾水份，放入冰鎮一下。

3. 調埋盆內放入醬料和洋蔥絲、芹菜段、紅辣椒段、蔥段拌勻，加入燙熟的海鮮拌勻，盛入盤內，放上小蕃茄、撒上香菜。

Ingredients

1 squid, 10 grass shrimp, 8 mussels, 1/2 onion (shredded), 4 strings celery (cut into sections), 2 red chili peppers(cut diagonally), 2 scallions (cut into sections), 6 cherry tomatoes (cut into halves), 2T minced cilantro

Seasonings

Combine: 2 Thai chili paste, 1T fish sauce, 1T tamarind, 2 1/2T lemon juice, 2/3T white granulated sugar, 2T minced garlic, 3 cloves shallots (shredded).

Methods

1. Rinse squid well, remove thin membrane and cut into slices. Remove heads from shrimps and retain tails, then devein and rinse well.

2. Blanch squid, shrimps and mussels in boiling water until done (in a Durotherm no liquid: cover and cook over medium heat until smoking, reduce to low and continue simmering for about 3 minutes). Remove and drain well, then let sit in ice water for some time.

3. Combine sauce ingredients with shredded onions, celery sections, red chili pepper sections and scallion sections well, then add blanced seafood and mix well. Remove to serving plate and top with cherry tomatoes, then sprinkle with cilantro. Serve.

Ingredients

300g chicken meat (shredded with hands), 2 Chinese cucumbers (diced), 150g cherry tomatoes (diced), 1 yellow bell pepper (diced), 300g iceberg lettuce leaves, 1 1/2T shredded young ginger, 2T minced scallion, 3T minced cilantro, 4T minced celery

Seasonings

4T soy sauce, 2T apple vinegar, 2T lemon juice, 1T sesame oil, 2 1/2T sesame paste

Methods

1. Combine shredded ginger, minced scallion, minced cilantro, and minced celery in a mixing bowl, add seasonings and mix well. Let rest for 5 minutes until the flavor is released.

2. Add shredded chicken, diced Chinese cucumbers, diced cherry tomatoes and diced yellow bell pepper to method (1). Mix until well blended.

3. Rinse lettuce leaves well, dry with vegetable dehydrator and line around the serving plate. Place the ingredients in center. Wrap the ingredients up with the lettuce leaves and serve.

義式
乳酪開胃沙拉
Italian Cheese Appetitzer Salad

以世界聞名的歌魯拱索拉藍黴起司為基底的沙拉醬，口感香濃滑順，
是一道相當有特色的沙拉，深得許多饕客喜愛，搭配白酒享用更是美味！

Dressings made with world famous Gorgonzola cheese have a smooth, thick texture.
Gorgonzola makes for a very special salad dish. It is loved by foodies the world over
and is especially good with white wine.

義式乳酪開胃沙拉

材料

雞胸肉300g.、馬鈴薯2個、牛蕃茄2個（每個切6瓣）、羅勒(九層塔)少許、洋蔥1/4個、西洋芹1/2支、櫻桃或藍莓12個、巴西里少許

起司醬汁

歌魯拱索拉藍黴起司120g.、無糖鮮奶油80g.、白酒2匙

做 法

1. 雞胸肉整塊鋪上西洋芹，放入蒸籠蒸約10分鐘(使用雙壁鍋：鍋中倒入3匙水，蓋上鍋蓋以中火煮至冒煙，轉小火續煮約7～8分鐘)。煮好後放涼，放入冰箱冷藏。

2. 馬鈴薯洗淨不去皮，以微波爐或電鍋煮熟(使用雙壁鍋：鍋中倒入1杯水(225c.c.)蓋上鍋蓋，以中火煮至冒煙，轉小火續煮約7～10分鐘，移外鍋燜7分鐘)，取出去皮切塊。

3. **製作醬汁**：將歌魯拱索拉起司和鮮奶油放入鍋內，以小火邊煮邊拌勻至成稠狀，淋上白酒拌勻。

4. 取出雞胸肉切片狀，與馬鈴薯、牛蕃茄一起盛盤，淋上起司醬汁，裝飾櫻桃（或藍莓）和巴西里。

完美烹調寶典
Tips for a Perfect Dish

1. 煮雞肉的時間需視肉的厚度自行調整。
2. 超級有味道的藍黴起司相當受到老饕歡迎，義大利的Gorgonzola是世界三大藍黴乳酪之一。另外兩個是法國的Roquefort和英國的Stilton。
3. 使用和裝置乳酪醬的器具都要擦拭乾淨（如菜刀、砧板、盤子等），才不容易發黴。
4. 用不完的起司以保鮮膜包好，放保鮮盒內，再入冰箱冷藏。

Ingredients

300g chicken breast, 2 potatoes, 2 beefsteak tomatoes (cut each into four equal wedges), basil as needed, 1/4 onion, 1/2 string celery, 12 cherries or blueberries, parsley as needed

Cheese Dressing

120g Gorgonzola cheese, 80g unsweetened whipping cream, 2T white wine

Methods

1. Line celery on the top of chicken breast and steam in steamer for 10 minutes (in a Durotherm: place chicken breast in pan with 3T of water, cover and steam over medium until smoking, reduce heat to low and continue cooking for about 7 to 8 minutes until done). Remove to cool and refrigerate

2. Rinse potatoes well and do not peel, cook in microwave or rice cooker until done (in a Durotherm: add 1 cup of water (225c.c.) to potatoes in pan, cover and cook over medium until smoking, then reduce heat to low and continue cooking for 7 to 8 minutes longer. Remove from the outer pan and let simmer for 7 minutes). Remove, peel off the skin, and cut into large pieces.

3. To prepare the dressing: combine Gorgonzola cheese and whipping cream together in pan, cook on low while stirring until thickenened, then drizzle with white wine to mix.

4. Remove chicken from the refrigerator and cut into slices. Transfer to a serving plate along with the potatoes and beefsteak tomatoes. Drizzle with the cheese sauce and garnish with cherries (or blueberries) and parsley. Serve.

Tips for a Perfect Dish

1. The cooking time of the chicken breast depends on the thickness of the chicken, adjust as needed.

2. Blue cheese with its superb taste is very popular among people who love food. Italian Gorgonzola is one of world's three famous blue cheeses, the other two being Roquefort (France) and Stilton (England).

3. The utensils for using and holding cheese have to be cleaned thoroughly (eg. kitchen knife, cutting board, plates, etc). to prevent mold.

4. The leftover cheese should be wrapped up well with saran wrap, kept in a tight container, and refrigerated.

加州風味
培根玉米
Californian Corn with Bacon

大人小孩都愛吃的美式風格料理。
An American style cuisine that both kids
and adults love.

加州風味培根玉米

材 料

黃色玉米3條、培根絲4匙(烤酥脆)、蒜末1 1/2匙、
巧達起司絲4匙、蝦夷蔥1匙、融化奶油1 1/2匙

調味料
黑胡椒、鹽適量

做 法

1. 玉米用玉米刨刀刨成玉米粒(要料理時再刨才
   能保有鮮甜味)。
2. 中火熱鍋,加入融化奶油,放入蒜末炒香,
   倒入玉米粒以小火炒至熟,如太乾加3匙高湯
   (使用雙壁鍋、休閒鍋:加入奶油,放入蒜末
   炒香蓋上鍋蓋以中火煮至冒煙,轉小火續煮約
   4分鐘),放入黑胡椒和鹽調味。最後加入培
   根絲、起司拌一下,盛盤時撒上蝦夷蔥。

完美烹調寶典
Tips for a Perfect Dish

1. 如無蝦夷蔥可改用一般較細的蔥花,或巴
   西里。
2. 玉米如買回家沒當天料理,需將玉米連同
   外皮一起放冰箱冷凍,才能保持玉米的鮮
   甜味。
3. 要將培根絲料理得脆酥,可以烤箱烤或入
   鍋煎炒的方式,超市也有罐頭包裝的酥脆
   培根絲。

Ingredients
3 ears yellow corn, 4T bacon strips
(baked until crispy), 1 1/2T minced
garlic, 4T shredded cheddar cheese,
1T chives, 1 1/2T melt butter

Seasonings
black pepper and salt as needed

Methods

1. Remove kernels from corn with a
   corn cutter (remove right before
   cooking to preserve freshness and
   sweetness).
2. Heat pan over medium heat, add
   melt butter until smoking, stir-fry
   minced garlic until fragrant, pour in
   corn kernels and stir over low heat
   until done. If the pan is too dry, add
   3T of soup broth (in a Durotherm
   or hot pan: add melt butter until
   smoking, stir-fry minced garlic until
   fragrant, add corn kernels and cover.
   Cook on medium until smoking,
   reduce heat to low and continue
   cooking for about 4 minutes more).
   Season with black pepper and salt
   to taste. Add shredded bacon and
   cheese to mix. Remove to a serving
   plate and sprinkle with chives.

Tips for a Perfect Dish

1. If chives are not available, use
   ordinary thin minced scallions or
   parsley.
2. Refrigerate the uncooked ears
   of corn with the husks on after
   purchasing from the market to
   preserve its sweetness.
3. To prepare crispy bacon strips, bake
   in oven or fry in pan. Supermarkets
   also carry canned crispy bacon
   strips.

越南風味豆腐
Vietnamese Style Tofu

只要是CC老師喜歡吃的越南菜，

CC老師就會將它改成大家比較能接受的口味，

這個略做調整的配方相當受到學生的歡迎喔！

CC teacher always changes her favorite dishes into dishes most people can accept. These ingredients have been adjusted and are quite popular with my students!

越南風味豆腐

材 料

四方油炸豆腐6塊、豬絞肉4匙、雞胸絞肉4匙、蝦仁4匙(切丁)、蒜末1匙、香菜梗末1匙、綠韭菜末4匙、香菜末2匙

調味料

淡色醬油1/2匙、白胡椒粉1/2小匙、鹽少許

辣醬魚露

白砂糖5匙、魚露2 1/2匙、檸檬汁3匙、蒜末2匙、紅辣椒末2匙

做 法

1. 油炸豆腐中間挖凹洞,放入烤箱中,以上下火200℃烤約5分鐘。豆腐挖出來的部分切碎備用。

2. 中火熱鍋,倒入1匙油,油熱後放入蒜末、香菜梗末爆香,先放入豬絞肉炒散、再加入雞絞肉和挖出來豆腐碎的部分炒勻,放入調味料炒熟後隨即加入蝦丁炒至蝦熟,最後倒入韭菜略炒幾下,即為餡料。

3. **製作辣醬魚露**:白砂糖加100c.c.熱水攪勻,待涼後與其他材料拌勻即成。

4. 將烤好的豆腐盛盤,餡料塞入豆腐凹洞中,淋上辣醬魚露、撒上香菜末即可食用。

完美烹調寶典
Tips for a Perfect Dish

1. 沒有烤箱也可以將豆腐放入180℃油鍋炸一下。

2. 辣醬魚露相當可口,可一次多做一些放在冰箱儲存,用來當做越式涼拌菜的醬汁或沾醬使用。

Ingredients

6 pieces square deep-fried tofu, 4T ground pork, 4T ground chicken breast, 4T shelled shrimp (diced), 1T minced garlic, 1T minced cilantro stems, 4T shredded green leek, 2T minced cilantro

Seasonings

1/2l light soy sauce, 1/2t white pepper, a pinch of salt

Spicy Fish Sauce

5T white granulated sugar, 2 1/2T fish sauce, 3T lemon juice, 2T minced garlic, 2T minced red chili pepper

Methods

1. Dig a hole in the center of the deep-fried tofu, place in oven and bake at 200℃ on the upper and lower elment for about 5 minutes, then remove. Chop the discarded tofu finely.

2. To prepare stuffing: Heat pan over medium heat and add 1T of oil until smoking, add minced garlic and cilantro stems and stir until fragrant. Stir-fry ground pork until separated, add ground chicken breast and chopped tofu. Stir quickly until evenly mixed, then season with seasonings to taste. Add diced shrimp and cook until done, then add leek to mix.

3. To prepare spicy fish sauce: add sugar to 100c.c. of hot water and stir until dissolved. Wait until cold and mix well with the rest of ingredients.

4. 4. To fill the stuffing into the tofu until all the tofu are done and place tofu to a serving plate and sprinkle the remaining stuffing on the side. Drizzle with spicy fish sauce and sprinkle with minced cilantro. It's ready to serve.

Tips for a Perfect Dish

1. If oven is not available, fry square tofu in oil at 180℃ for a minute and remove.

2. Spicy fish sauce is delicious. It can be prepared in large quantity and kept in the refrigerator. It can be used as drizzing sauce or dipping sauce for Vietnamese cold dishes.

泰式月亮蝦餅
Thai Shrimp Moon Cake

必點必學！這道泰國佳餚是所有學生的最愛。

Must-order and must-learn dish!
This delicious Thai dish is loved by all my students.

泰式月亮蝦餅

材料

蝦仁300g.、豬絞肉100g.(帶肥)、春卷皮4張、香菜末2匙、荸薺8粒、玉米粉2匙、蛋1個

調味料

白胡椒粉1/2小匙、魚露1/2小匙、鹽1/3小匙

沾醬

魚露2匙、白砂糖4匙、檸檬汁2匙、蒜末1/2匙、紅辣椒末1/2匙、香菜末1/2匙

做法

1. 蝦仁切末，荸薺拍碎去掉水份。放入調理盆內，加入絞肉、香菜末、蛋、玉米粉及調味料拌勻成餡料。
2. 取一張春卷皮攤平，鋪上1/2的餡料，稍壓平再蓋上一張春卷皮，用叉子叉幾下。一共做2份。
3. 放入約160℃油鍋中，炸至呈金黃色，取出瀝乾油，切成三角形狀。沾醬拌勻，放入小碟中，上桌時一併附上。

完美烹調寶典
Tips for a Perfect Dish

1. 食用時搭配沾醬，萵苣生菜及涼拌大黃瓜更是美味無比。
2. 油炸食物時需待所有食物取出再熄火，以免食物因溫度驟降吸附更多油。
3. 萵苣生菜也叫美生菜、結球萵苣，在一般超市都有賣。

CC老師附贈一道好搭配小菜

大黃瓜涼拌做法：

1/4條大黃瓜切片+紅蔥頭絲3顆+白醋3匙+白砂糖1匙+鹽1/2匙，拌勻放置10分鐘即可享用。

Ingredients

300g shelled shrimp, 100g ground pork (with fat on), 4 springroll wrappers, 2T minced cilantro, 8 water chestnuts, 2T corn flour, 1 egg

Seasonings

1/2t white pepper, 1/2t fish sauce, 1/3t salt

Dippping Sauce

2T fish sauce, 4T white granulated sugar, 2T lemon juice, 1/2T minced garlic, 1/2T minced red chili pepper, 1/2t minced cilantro

Methods

1. Mince the shrimp. Crush water chestnuts to remove the liquid. Place shrimp and water chestnuts in a mixing bowl, along with ground pork, minced cilantro, egg, corn flour and seasonings added. Stir until the mixture is evenly mixed.
2. Spread a spring roll wrapper evenly out on the surface, spread half of the mixture from method (1) evenly over on the wrapper, press gently and cover with another wrapper. Prick holes with a fork to let the air through. Make two portions in total.
3. Fry in oil at 160℃ until golden and remove to drain. Cut into triangular wedges. Combine the seasonings to make the dipping sauce. Transfer to a saucer dish and serve on the side of the moon cake.

Tips for a Perfect Dish

1. It will taste even better if dipping sauce, lettuce leaves and a cold cucumber dish are served on the side.
2. When deep-frying, remove all the ingredients from oil first before turning off the heat, to prevent the ingredients from absorbing more oil after the temperature drops.
3. American iceberg lettuce can be found in most supermarkets, even in traditional markets.

An extra side dish as a gift
Cold Cucumber Dish

Combine 1/4 sliced cucumber, 3 shredded shallots with 3 tablespoons of white vinegar, 1 tablespoon of white granulated sugar and 1/2 tablespoon of salt together, stir until evenly mixed and let sit for at least 10 minutes before serving.

西班牙
馬鈴薯蛋餅
Spainish Potato Egg Omelet

不知不覺一直吃，真是恐怖的美味！

CC的身材不是一天造成的，而是美食這兇手！

Just keep eating without thinking about it – what a horrifying experience!
CC's figure is not built in a day: the delicious cuisine is what caused it!

西班牙馬鈴薯蛋餅

材 料

馬鈴薯600g.、培根5片(切絲)、洋蔥1個(切末)、蛋6個、無糖鮮奶油1/3杯、融化奶油2匙、羅勒(九層塔，切末)1/4杯、黑胡椒1/2匙、鹽適量、巴西里適量

做 法

1. 馬鈴薯放入鍋內蒸熟去皮切片。
2. 中火熱鍋，放入11/2匙橄欖油，待油熱後(油紋產生)放入培根煎成酥脆狀取出，倒入洋蔥末炒透(呈淡茶褐色)。
3. 蛋去殼倒入調埋盆內打勻，加入鮮奶油和做法2.的培根、洋蔥和羅勒，加入黑胡椒和鹽拌勻，再倒入蒸熟的馬鈴薯片和奶油拌勻。
4. 中火熱鍋，放入11/2匙橄欖油，待油熱倒入做法3.全部材料，轉小火煮約15～20分鐘(使用雙壁鍋：蓋上鍋蓋轉小火煮約10分鐘即可)，再移入烤箱上下火200℃烤至呈金黃色，最後撒上巴西里。

Ingredients

600g potatoes, 5 strips bacons (shredded), 1 onion (minced), 6 eggs, 1/3C unsweetened whipping cream, 2T melt butter, 1/4 cups sweet basil or basil (minced), 1/2T black pepper, a pinch of salt, parsley as needed

Methods

1. Steam potatoes in pan until done, remove the skin and cut into slices.
2. Heat pan over medium, add 1 1/2T of olive oil until smoking, fry bacon until crispy and remove. Use the remaining oil to saute minced onion until done through (light brown).
3. Remove shell from egg and beat well in mixing bowl, add whipping cream and bacon and basil from method (2) as well as basil, season with black pepper and salt to taste. Then pour into potato slices and mix well with butter.
4. Heat pan over medium, add 1 1/2T of olive oil until smoking, pour in method (3) and reduce heat to low. Cook for about 15 to 20 minutes (in a Durotherm: cover and reduce to low, cook for 10 minutes until done). Remove to oven at 200℃ on upper and lower element until golden, then sprinkle with parsley. Serve.

歐風馬鈴薯派
European Potato Pie

這道料理太受歡迎了,曾被學生取用在蘋果日報上示範,
做法簡單、口味濃郁,居家小點心或宴客菜都適宜。

This dish is so popular, that it was once demonstrated on Apple
Daily by the students. The methods are simple and flavor is
marvelous. It can be homemade as a snack or a banquet dish.

歐風馬鈴薯派

材 料

馬鈴薯600g.(去皮切片)、培根1/2杯(切絲)、洋蔥末1杯、低筋麵粉1/2杯(過篩)、蛋3個、融化奶油3匙、雞湯粉1匙、莫茲瑞拉起司絲150g.、黑胡椒1/2匙、巴西里適量、帕瑪森起司粉3匙、無糖鮮奶油4匙

做 法

1. 馬鈴薯片放入蒸鍋蒸爛,搗成泥狀。
2. 取2匙奶油放入鍋內以極小火煮至融化,加入培根絲以小火炒至呈焦黃色,接著倒入洋蔥末炒透。
3. 麵粉倒入馬鈴薯泥中,加入蛋、1匙奶油、雞湯粉和70g.莫茲瑞拉起司絲、黑胡椒、鮮奶油,以及做法2.拌勻。
4. 使用烤箱:做法3.放入烤皿內,移入烤箱、以上下火140～150℃烤約40分鐘取出,鋪上剩下的莫茲瑞拉起司絲和帕瑪森起司粉,烤至呈金黃色即成,最後撒上巴西里。
5. 使用雙壁鍋或休閒鍋、材質好的平底鍋:取一張烤焙紙鋪在鍋內,倒入做法3.,均勻撒上剩下的莫茲瑞拉起司絲和帕瑪森起司粉,蓋上鍋蓋以小火烤約30分鐘,最後撒上巴西里。

Ingredients

600g potato (peeled and sliced), 1/2C bacon (shredded), 1C minced onoin, 1/2C cake flour (sifted), 3 eggs, 3T melt butter, 1T chicken soup broth powder, 150g shredded mozzarella cheese, 1/2T black pepper, parsley as needed, parmesan cheese powder, 4T unsweetened whipping cream

Methods

1. Steam potato slices in steamer until soft and crumbling, then mash well.
2. Melt 2T of butter in pan over low heat, add bacon and saute over low heat until yellow brown, then add minced onion and fry until done through.
3. Add flour in the mashed potato along with eggs, 1T of butter, chicken soup broth powder, 70g of mozellar cheese, black pepper, whipping cream and method (2).
4. Use an oven: Place method (3) in a baking bowl, bake in oven at 140 to 150℃ on upper and lower element for about 40 minutes, remove and spread evenly with remaining mozellara cheese and parmessan cheese. Return to oven and bake until golden and remove. Sprinkle with parsley.
5. In a Durotherm, hot pan or a better quality pan: Line a sheet of parchment paper on bottom of frying pan, pour in method (3) and sprinkle evenly with remaining mozellara cheese and parmessan cheese. Cover and fry over low heat for about 30 minutes, then sprinkle with parsley. Ready to seve.

義大利蛋餅
Italian Omelet

這道健康又好吃的歐姆雷特，看起來就讓人忍不住想咬一口；

是義大利媽媽常做的家庭料理，早、午、晚餐都適合喔！

This healthy and delicious omelet. You can't help but take a bite! It is a popular home-made cuisine prepared by mothers across Italy. It is good for breakfast, lunch, or dinner.

義大利蛋餅

材料

蘆筍100g.(斜切約1公分段)、蘑菇罐頭1/2罐(去水、切片)、洋蔥1個(切絲)、橄欖油2大匙、紅黃甜椒各1/2個(切丁)、牛蕃茄1個(切丁)、火腿100g.(切丁)、蒜末1匙、帕瑪森起司粉4匙、蛋8個、奶油3匙、黑胡椒、鹽滴量

做法

1. 奶油放入鍋內以極小火煮至融化,加入洋蔥絲以小火炒透,隨即倒入蘆筍、蘑菇炒約1分鐘,起鍋放涼備用。
2. 蛋去殼,放入調理盆內,倒入放涼了的做法1.,加入牛蕃茄丁、彩椒丁、蒜末和火腿丁,撒入起司粉、黑胡椒和鹽,拌勻。
3. 使用烤箱‧做法2.放入烤皿內,移入烤箱以上下火160℃烤約20分鐘,或使用一般鍋烤薄些,即放入的量少一些,如此較容易烤熟。
4. 使用雙壁鍋,休閒鍋或材質好的平底鍋:鍋中倒入2匙橄欖油,先將鍋面塗抹一層油,開中火,待油熱後(產生油紋)加入做法2.,蓋上鍋蓋改小火煎約10分鐘,熄火,移入外鍋燜4分鐘。

完美烹調寶典
Tips for a Perfect Dish

1. 帕瑪森起司和火腿本身就有鹹味,放鹽調味時需留意鹹度而調整份量。
2. 帕瑪森起司可買整塊的,要料理時再磨粉較香。

Ingredients

100g asparagus (cut diagonally into 1cm long slices), 1/2 can canned mushrooms (water removed and sliced), 1 onion (shredded), 2T olive oil, 1/2 red and yellow bell pepper each (diced), 1 beefsteak tomato (diced), 100g ham (diced), 1T minced garlic, 4T parmesan cheese powder, 8 eggs, 3T butter, black pepper and salt as needed

Methods

1. Melt butter in pan over low heat until done, add shredded onion and stir-fry over low heat until thoroughly cooked, Add asparagus and mushrooms, stir for about 1 minute and remove from heat to cool.
2. Break eggs into the mixing bowl, return cool method (1) along with diced beefsteak tomato, bell peppers, minced garlic and ham. Sprinkle with cheese powder, black pepper and salt to taste. Combine until evenly mixed.
3. Use oven: place method (2) in a baking bowl, and bake in oven at 160℃ on the upper and lower element for about 20 minutes. Or use ordinary thin pan and line with the ingredients in an even, thin layer. It is easier and quicker to bake.
4. In a Durotherm, hot pan or good quality frying pan: add 2T of olive oil in pan to grease the whole surface, reduce heat to medium until the oil is smoking, add method (2) and spread evenly. Cover and reduce heat to low and fry for about 10 minutes, remove from outer pan and simmer for 4 minutes until done. Ready to serve.

Tips for a Perfect Dish

1. Parmesan cheese and ham are seasoned with salt already, adjust the portion of the salt when seasoning.
2. For a better and stronger flavor, purchase the whole chunk of parmesan cheese and grate right before cooking.

在炎炎夏天吃海鮮是連神仙都羨慕的事情吧！
無論是東南亞那酸酸甜甜的魚蝦，
還是日本那清清爽爽的蒸魚；
無論是地中海旁鮮活跳躍的海鮮，
還是法國、義大利小城的燴海鮮……
都是滿滿一整個的美味，滿滿一整盤的歡欣。

It'll fire your imagination even the Deity
if you can enjoy the seafood in the hot summer.
Whether is the seafood with the sour-and-sweet dressing
or just light-taste of steam fish,
And whether is the live and fresh seafood of Med
or braised seafood of France/Italy villages,
Every dish is full filled with delicacy、full filled with cheer.

最想學會的外國菜

Part2 海鮮類

Seafood Section

Sugarcane shrimp is the most representative dish of Vietnamese cuisine. Seldom do restaurants cook it well and delicious versions are not cheap either. Therefore, this dish is also CC students' favorite. One of the must-order and must-learn dishes.

越南甘蔗蝦
Vietnamese Sugarcane Shrimp

甘蔗蝦是越南菜最具代表的料理，
做得好吃的店家很少，好吃的又不便宜；
所以這道料理也是CC學生的最愛，必點必學的菜色之一。

越南甘蔗蝦

材 料

蝦仁400g.、蛋白1個、白膘20g.(白色的豬肥肉)、玉米粉2匙、香菜末2匙、白砂糖1/2小匙、鹽1/2小匙、甘蔗14支(約12×1.2公分)、萵苣生菜200g.、酒1/2匙、新鮮薄荷適量

沾 醬

白砂糖2 1/2匙+水4匙煮沸熄火待涼+檸檬汁1 1/2匙、魚露1 1/2匙+蒜末1 1/2匙+紅辣椒末1匙拌勻

做 法

1. 將沾醬的白砂糖加4匙水煮沸,待涼後與其他材料拌勻。
2. 蝦仁去腸泥、拭乾水份,剁碎放入調理盆內;倒入蛋白、白膘、玉米粉、香菜末、鹽、白砂糖和酒,拌至有黏性。
3. 分成14等份、做成圓球狀,一一包住甘蔗中段,上下段各留約2公分,表面可塗上少許油使之光滑。
4. 放入160℃油鍋炸至呈金黃色,取出瀝乾油置盤中,食用時附上沾醬、搭配生菜和薄荷食用。

完美烹調寶典
Tips for a Perfect Dish

1. 沒有新鮮薄荷無妨,只搭配生菜也很美味。
2. 蝦仁加了白膘,混合起來口感才不會澀。

Ingredients

400g shelled shrimp, 1 egg white, 20g white pork fat, 2T corn flour, 2T minced cilantro, 1/2t white granulated sugar, 1/2t salt, 14 sugar cane sections (approximately 12x1.2cm long), 200g iceberg lettuce leaves, 1/2T cooking wine, fresh mint leaves as needed

Dipping Sauce

Combine2 1/2 tablespoons of white granulated sugar with 4 tablespoons of water, cook until sugar dissolved, then set aside to cool, then add 1 1/2 tablespoons of lemon juice, 1 1/2 tablespoons of fish sauce and 1 tablespoon of minced red chili pepper. Stir to mix.

Methods

1. Combine white granulated sugar with 4 tablespoons of water together, bring to a boil and then let cool, then add the rest of the ingredients.
2. Devein shrimp and dry, then chop finely and remove to a mixing bowl. Add egg white, pork fat, corn flour, minced cilantro, salt, white granulated sugar and wine. Keep stirring until it is sticky.
3. Divide into 14 equal portions, roll each portion into a round ball and wrap around the middle section of each sugar cane, leaving 2cm on the top and bottom. Brush the surface with a little oil to make it smooth.
4. Deep-fry in oil at 160℃ until golden, remove to drain and transfer to a serving plate. Serve the dipping sauce, lettuce leaves and mint leaves on the side.

Tips for a Perfect Dish

1. It is OK if mint leaves are not available, the dish is still good with iceberg lettuce only.
2. The shrimp will not taste as dry if pork fat is added.

清蒸檸檬魚
Steamed Lemon Fish

這是CC
老師的招牌
教學菜之一，
進出泰國近30
次，吃遍各大小餐
廳的料理方式，這個
配方是CC最愛的味道，
請大家嚐嚐。

This dish is one that CC teacher
is an expert on. CC teacher has
been to Thailand almost 30 times
and been to almost every large and
small restaurant. This recipe is CC's
favorite and she would like to share it
with everybody.

清蒸檸檬魚

材 料
鱸魚1條約800g.、蒜末3大匙、香菜1/4杯、紅辣椒末11/2匙

調味料
南薑末1匙、檸檬汁31/2匙(依個人喜愛酸度調整)、高湯1/4杯、白砂糖1匙、魚露2匙

做 法
1. 香菜洗淨,分成三部分,頭留住、葉子和莖部分開,莖切細末。
2. 魚拭乾正反面,各撒上1/2小匙鹽,放盤內稍醃一下。放入蒸籠,將香菜頭鋪在魚上(約3個香菜頭)蒸12分鐘左右後取出,放盤上,再鋪上香菜葉。

葉子

/莖部

香菜頭

3. 調味料拌勻,加入蒜末、香菜莖末、南薑末和紅辣椒末,全部煮沸後淋在魚身上即成。若家中有不鏽鋼蒸魚器皿,可直接將調味醬汁混勻淋在魚上一起蒸,味道才會更好吃。
4. CC選用的休閒鍋可將魚直接放入鍋內,鋪上香菜頭、倒入3匙高湯、蓋上鍋蓋,冒煙後轉小火煮約6分鐘,淋上調好的醬汁再煮1分鐘熄火,移至外鍋放置約6分鐘,打開鍋蓋撒上香菜即可享用。此種料理方式較不擔心魚蒸過老、食用時魚肉較嫩及甜美,還可直接上桌,保持熱度。

完美烹調寶典
Tips for a Perfect Dish

1. 魚露的量可參考CC的比例再慢慢隨自己口味調整鹹度。
2. 南薑放於冷凍庫可保存2年。

Ingredients
1 sea bass (about 800g), 3T minced garlic, 1/4C cilantro, 1 1/2T minced red chili pepper

Seasonings
1T minced tumeric, 3 1/2T lemon juice (adjust as desired), 1/4C soup broth, 1T white granulated sugar, 2T fish sauce

Methods
1. Rinse cilantro well, divide into three parts. Retain the roots, separate leaves from stems, then mince stems finely .
2. Dry fish well on both sides, sprinkle with 1/2t of salt on each side, then marinate for some time. Remove to steamer, spread cilantro roots over fish evenly (approximately 3 cilantro roots). Steam for about 12 minutes and remove. Transfer to a serving plate and garnish with cilantro leaves.
3. Combine the seasonings well, add minced garlic and minced cilantro stems, tumeric and red chili pepper. Mix well and heat until boiling in pan, remove and drizzle over fish. If steaming plate is available, combine seasonings and drizzle directly over fish, then steam together, the flavor will be thoroughly absorbed.
4. CC uses a hot pan, which allows the fish be cooked directly inside, then spreads cilantro roots on the fish and adds 3 tablespoons of soup broth. Cover and cook until steam comes out, then reduce heat to low and cook for about 6 minutes. Drizzle with well-mixed seasonings and continue cooking for 1 more minutes. Remove from heat and let sit for 6 minutes before opening the top, then sprinkle with cilantro and serve. This preparation prevents the fish from getting too too tough from over cooking. The fish meat will be more tender and sweeter. It may also be served directly to the dining table to maintain heat.

Tips for a Perfect Dish
1. To prepare the fish sauce, see CC's proportion as a reference and adjust as needed.
2. Turmeric can be frozen for as long as two years.

北海道蒸鱈魚
Hokkaido Steamed Cod

很特別的料理方式，

健康味美，深得學生們強力推薦，

尤其年齡稍長者特別喜愛，口感方面也很順滑。

This is a very special cooking method, healthy and delicious. Strongly recommended by the students, especially the older ones. The texture is very smooth and slick.

北海道蒸鱈魚

材 料
鱈魚2片、紅蘿蔔30g.(切絲)、杏飽菇1條(切絲)、新鮮香菇5朵(切絲)、金針菇1/2包、蔥4支(2支切段、2支切絲)、蛋白3個、紅辣椒1支(切絲)

日式高湯
柴魚精1匙、水8匙

調味醬汁
醬油3匙(淡味)、日式高湯4匙、酒2匙、味醂1/2匙

做 法
1. 先拌勻1匙柴魚精和8匙水調成8匙日式高湯。其中4匙與其他調味醬汁材料放入鍋中煮沸。
2. 將蔥段鋪於盤中，放上鱈魚，淋上4匙日式高湯，整盤沸水蒸籠蒸約10分鐘(使用雙壁鍋或休閒鍋；蓋上鍋蓋，以中火煮至冒煙轉小火煮4分鐘即成)。
3. 蛋白打至發泡，放入紅蘿蔔絲、杏鮑菇絲、香菇絲、金針菇，和蔥絲拌勻。
4. 做法3.鋪於做法2.上，再蒸煮約6分鐘（使用雙壁鍋或休閒鍋：只要蒸4分鐘），最後再撒上紅辣椒絲、蔥絲，淋上調味醬汁即可享用。

Ingredients
2 slices cod, 30g carrot (shredded), 1 oyster mushroom (shredded), 5 fresh shiitake mushrooms (shredded), 1/2 pack Enokitake mushrooms, 4 scallions (2 cut into sections, 2 shredded), 3 egg whites, 1 red chili pepper (shredded)

Japanese Soup Broth
1T bonito powder, 8T water

Sauce
3T soy sauce (light), 4T Japanese soup broth, 2T cooking wine, 1/2T mirin

Methods
1. Combine 1 tablespoon of bonito powder and 8 tablespoons of water to make Japanese soup broth. Remove 4 tablespoons of broth to mix with the sauce. Bring to a boil in a pan.
2. Spread scallion sections in steaming plate and top with cod. Drizzle with 4 tablespoons of soup broth, remove to a steamer with boiling water inside and steam for about 6 minutes (use a Durotherm or hot pan: cover and cook over medium until steam comes out, reduce heat to low and continue cooking for 4 minutes).
3. Beat egg whites until fluffy, add shredded carrot, oyster mushrooms, shiitake mushrooms, and enokitake mushrooms as well as shredded scallion to mix.
4. Spread method (3)over on method (2), steam for another 6 minutes. (Use a Durotherm or hot pan: steam for 4 minutes only). Sprinkle with shredded red chili pepper, and shredded scallion, then drizzle with sauce. Serve.

Tips for a Perfect Dish
1. After the scallions and red chili pepper are shredded, soak in cold water to remove any muddy flavor and increase their crunchiness.
2. It is very easy to prepare Japanese soup broth. Just combine 1 tablespoon of bonito powder with 8 tablespoons of water evenly.

愛蒙特起司
鮮魚
Emmentaler Cheese Fish

愛蒙特起司是瑞士非常有名的起司，
味道香濃，用在料理上特別好吃。
這個起司的特色就是一個小洞一個小洞的，
卡通影片裡小老鼠愛偷吃的起司都長這個樣子。

Emmentaler is a very famous Swiss cheese with a thick aroma. It is especially good in cooking. This cheese is special because of its holes, just like the cheese that Jerry loves in the Tom and Jerry cartoons.

愛蒙特起司鮮魚

材料

白魚肉或鯛魚肉400g.(去皮去骨)、愛蒙特起司100g.、牛蕃茄1個(去皮切丁)、黃紅甜椒各1/2個(切菱型丁狀)、馬鈴薯11/2個、牛奶1/4杯、白酒80c.c.、黑胡椒和鹽適量

做 法

1. 將黑胡椒和鹽以2：1比例調配，均勻塗在魚的正反面醃一下。
2. 馬鈴薯去皮蒸熟，放入調理盆內搗成泥狀，加入牛奶和少許鹽拌勻。
3. 牛蕃茄和彩椒放入調理盆內，加入適量的黑胡椒、鹽和1/2匙橄欖油拌勻。
4. 中火熱鍋，倒入2匙橄欖油，待油熱後(油紋產生)，放入魚煎兩面各煎約3分鐘（需視肉的厚薄調整），淋上白酒熄火，鋪上馬鈴薯泥，再放上愛蒙特起司、撒上做法3.，移入烤箱以上下火230℃烤至呈金黃色即成。

完美烹調寶典
Tips for a Perfect Dish

1. 家裡沒有烤箱，可在撒上做法3.時，轉小火、蓋上鍋蓋，待起司融化即可（如果是一般鍋子要特別小心鍋底燒焦，火候一定要控制好）。
2. 魚也可以雞胸肉替代。使用休閒鍋則時間可再縮短

Ingredients

400g white fish meat or seabream (peeled and boned), 100g Emmentaler cheese, 1 beefsteak tomato (peeled and diced), 1/2 yellow and red bell peppe each (cut into diamond shape pieces), 1 1/2 potato, 1/4C milk, 80c.c. white wine, black pepper and salt as needed

Methods

1. Prepare black pepper salt 2:1 and spread evely over on both sides of the fish. Sit for a while.
2. Peel potato and steam until done, mash in a mixing bowl and mix well with milk and a little salt to taste.
3. Add beefsteak tomato and bell peppers to the bowl, pour in black pepper and salt along with 1/2T of olive oil to mix.
4. Heat pan over medium, add 2T of olive oil until smoking, fry fish on both sides for about 3 minutes. (Depends on the thickness of the fish meat). Drizzle with white wine and remove from heat. Spread mashed potato over top evenly, then top with cheese and sprinkle with method (3). Remove to oven and bake at 230℃ on the upper and lower element until golden. Ready to serve.

Tips for a Perfect Dish

1. If an oven is not available, when sprinkle with method (3), reduce heat to low, then cover until cheese melts (if ordinary pan is used, control the heat well to prevent the bottom from getting burnt).
2. Chicken breast meat may be used instead of the fish.

地中海味
炸海鮮
Mediterranean Deep-Fried Seafood

清新的風味綜合了香料做成的麵衣，產生的香味令人著迷。

A light clear flavor with an herbal coating. Its aroma is entirely fascinating.

地中海味炸海鮮

材料
蝦12隻、透抽200g.、鯛魚200g.

麵衣
低筋麵粉1杯(過篩)、小蘇打粉1小匙(過篩)、蒜末1匙、羅勒末2匙(九層塔)、柳丁皮末1/2匙、檸檬皮末1/2匙、橄欖油1匙、蛋2個

香料鹽
迷迭香末1匙、羅勒末2匙、百里香末1/2匙、蒜末1匙、柳丁皮末1/2匙、檸檬皮末1匙、鹽1匙、黑胡椒粉2匙，拌勻

做法
1. 蝦去殼留尾巴，由背部劃開去腸泥，透抽去薄膜洗淨瀝乾水份、切圈狀，鯛魚切成長約6公分、寬約1公分。香料鹽材料拌均勻備用。
2. 製作麵衣：低筋麵粉和小蘇打粉過篩，加入全部材料和3/4杯水拌勻成稠狀。
3. 將海鮮料沾裹麵衣，放入160℃油鍋，炸至呈金黃色，取出瀝乾油放入盤中，上桌時附上香料鹽。

完美烹調寶典 *Tips for a Perfect Dish*
1. 用鋼材好的平底鍋油炸食物，可省掉很多油，CC常用的鍋為寬24公分平底鍋，油炸食物時，只要倒約1公分高的油，就可以將食物炸得酥脆。
2. 也可將南瓜切片或地瓜切片沾上麵衣炸，超好吃哦！
3. 油炸食物好吃的關鍵在於油溫，必須了解各種食用油的燃點至幾度才好選用。

常用油的燃點和適合烹飪的方法

油脂種類	油燃點	適合的烹飪法	成分
大豆沙拉油	160 ℃	水炒、中火炒	飽和脂肪酸
橄欖油	180 ℃	涼拌、水炒 (Extra Virgin Olive Oil)	單元性不飽和脂肪酸
	200 ℃	中火炒 (Pure Oil)	
葵花油	210 ℃	涼拌、水炒	多元性不飽和脂肪酸

Ingredients
12 shrimps, 200g squids, 200g seabream

Coating
1C cake flour (sifted), 1t baking soda (sifted), 1T minced garlic, 2T minced sweet basil (basil), 1/2T orange zest, 1/2T lemon zest, 1T olive oil, 2 eggs.

Flavored Salt
1T minced rosemary, 2T minced basil, 1/2T minced thyme, 1T minced garlic, 1/2T orange peel, 1T minced lemon peel , 1T salt, 2T black pepper

Methods
1. Remove shell from shrimp and retain the tails, then dovein and rinse. Remove membrane from squids, rinse well and drain, then cut into round sections. Cut bream into pieces about 6cm long and 1cm wide. Combine the ingredients from the flavored salt.
2. To prepare coating: Combine sifted cake flour and baking soda in a mixing bowl, add the remainder of the ingredients along with 3/4C of water. Mix well until thickened.
3. Coat seafood evenly with a layer of coating. Deep-fry in 160℃ oil until golden. Remove to drain, then transfer to a serving plate and serve with flavored salt on the side for dipping.

Tips for a Perfect Dish
1. Using a better quality stainless steel frying pan to deep-fry food. It can save much cooking oil. The frying pan that CC uses is 24cm wide, when deep-frying, only 1cm high of oil is needed to deep-fry food until crispy.
2. Pumpkin or yam can be sliced and coated with coating, then deep-fried. It is super good!
3. The key point to a delicious deep-fried dish is the temperature of the frying oil. You have to understand the boiling point of every single cooking oil type before you purchase.

這道菜是過年時最受歡迎的年菜料理之一，
當然不論宴客或開派對時也都是很精彩的菜色！
**This dish is one of the most popular dishes during New Year,
wonderful in banquets or parties.**

里昂燴鮮蝦
Leon Braised Shrimp

里昂燴鮮蝦

材料

大蝦或明蝦4隻、南瓜丁2匙、牛蕃茄丁2匙、蒜末1/2匙、洋蔥末1匙、蘑菇罐頭2匙(去水切丁)、大蒜奶油醬1/2匙(做法參考P.127)、巴西里少許、莫茲瑞拉起司4匙、白酒2匙、黑胡椒、鹽適量

做法

1. 中火熱鍋，放入1/2匙橄欖油，油熱後加入蒜末和洋蔥末炒香，放入南瓜丁炒熟，再加入蘑菇丁、蕃茄丁炒幾下，以黑胡椒、鹽調味。

2. 蝦由背部劃開、取出腸泥後剝開蝦肉，淋上白酒、撒上少許黑胡椒、鹽，塗上大蒜奶油。放入上下火220℃烤箱烤至蝦約5分熟。取出鋪上做法1.，再撒上起司，繼續放入烤箱烤至起司融化即可食用，上桌前撒上巴西里裝飾。

3. 使用休閒鍋：蝦由背部劃開、取出腸泥後剝開蝦肉，淋上白酒、撒上少許黑胡椒粉、鹽，塗上大蒜奶油，放入鍋內，蓋上鍋蓋以小火煮約3分鐘。鋪上做法2.後再蓋上鍋蓋，小火煮3分鐘讓起司融化，上桌前撒上巴西里裝飾。

完美烹調寶典
Tips for a Perfect Dish

1. 起司在一般超市都有賣，不限於莫茲瑞拉起司，也可選用自己喜愛的口味。
2. 開派對時可以較小隻的蝦取代大蝦或明蝦，比較小巧可愛。
3. 由於蝦本身具有甜味，故鹽不要放太多，且大蒜奶油及起司均帶有些鹹味。

Ingredients

4 large shrimps or prawns, 2T diced pumpkin, 2T diced beefsteak tomato, 1/2T minced garlic, 1T minced onion, 2T canned mushrooms (water removed and diced), 1/2T garlic flavored butter (see methods on p.127), parsley as needed, 4T mozzarella cheese, 2T white wine, black pepper and salt as needed

Methods

1. Heat pan over medium, add 1/2T of olive oil until smoking, then stir-fry minced garlic and onion until fragrant. Add diced pumpkin until done, add diced mushrooms and tomato to mix. Season with black pepper and salt to taste.

2. Devein shrimp and spread the shrimp wide, drizzle with white wine and sprinkle with black pepper and salt to taste, then brush with garlic flavor butter on top. Bake in oven at 220℃ on the upper and lower element until half done. Remove and spread with method (1), and sprinkle with cheese powder, then return to oven and continue baking until cheese melt. Remove and sprinkle with parsley to garnish before serving.

3. In a hot pan: Devein shrimp and spread shrimp wide, drizzle with white wine and sprinkle with black pepper and salt to taste, then brush with garlic flavor butter on surface. Return to pan and cover. Cook over low heat for about 3 minutes. Spread method (2) on top, then cover and continue cooking on low for 3 more minutes until cheese melts. Remove and sprinkle with parsley to garnish before serving.

Tips for a Perfect Dish

1. Cheese can purchased at ordinary supermarkets. Select any cheese as desired.
2. Use smaller shrimp instead of large ones or pawns to prepare party dishes as they look cuter.
3. Shrimp is sweet itself, do not add too much salt since the garlic flavor butter and cheese are already salty.

這道豐盛的海鮮料理非常具有義大利菜的風味，
示範時總是備受學生喜愛；可搭配法國麵包或拌上義大利麵
一起享用，尤其拿麵包沾著醬汁吃，彷彿置身人間天堂！

This rich, delicious seafood dish is very Italian. It is a student favorite
during demonstration. It can be served with French bread or a plate of
spagetti. When you dip the bread in the sauce, you feel like heaven!

帕多瓦燴海鮮
Padua Braised Seafood

帕多瓦燴海鮮

材料

鯛魚300g.、蝦8隻、透抽1隻、蕃茄粒罐頭1罐（搗碎）、蒜末1 1/2匙、洋蔥末3 1/2匙、麵包粉3匙、白酒100c.c.、高湯1/3杯（一般鍋3/4杯）、羅勒末6匙、牛蕃茄丁2/3杯、迷迭香末1匙、麵粉3匙、黑胡椒、鹽適量

做法

1. 鯛魚抹上黑胡椒、鹽(比例2：1)，兩面均勻沾上麵粉。
2. 蝦去殼留尾巴，背部劃一刀去除腸泥，透抽洗淨去薄膜、切圓圈狀。
3. 中火熱鍋，倒入2匙橄欖油，待油熱後（油紋產生）加入蒜末和洋蔥末炒透，放入鯛魚煎至兩面約八分熟。
4. 調理盆內倒入牛蕃茄丁、3 1/2匙羅勒末、迷迭香末和1/3匙鹽、1小匙黑胡椒拌勻。
5. 將鯛魚放入耐烤皿中，均勻鋪上做法4.，撒上麵包粉，入烤箱以上下火230°烤約5分鐘，取出盛盤。
6. 中火熱鍋，倒入1 1/2匙橄欖油，油熱後加入蒜末和洋蔥末炒透，放入蝦和透抽，撒上約1/3小匙的黑胡椒和鹽，拌炒幾下，淋上白酒，煮約30秒，放入蕃茄粒和高湯再煮一下，最後拌入羅勒末。整個澆在魚上即可享用。

完美烹調寶典
Tips for a Perfect Dish

1. 台灣的九層塔也屬於羅勒的品種之一，台灣的品種較容易發黑、但也較香、味道較重。
2. 乾燥的香料用量需少於新鮮香料一倍的量，這道菜使用的迷迭香建議選用新鮮的。

Ingredients

300g bream, 8 shrimps, 1 squid, 1 canned whole tomato (crushed), 1 1/2T minced garlic, 3 1/2T minced onion, 3T bread crumbs, 100c.c. white wine, 1/3C soup broth (measuring cup 3/4C), 6T minced sweet basil, 2/3C diced beefsteak tomato, 1T minced rosemary, 3T flour, black pepper and salt as needed

Methods

1. Coat sea bream with black pepper and salt (2:1), then flour evenly on both sides.
2. Remove shell from shrimp and retain the tails, devein and rinse well. Rinse squid well and remove the membrane, then cut into circular slices.
3. Heat pan on medium, add 2T of olive oil until smoking, stir fry minced garlic and onion until done, put in bream and fry until 80% done on both sides.
4. Combine diced beefsteak tomato, 3 1/2T minced basil, minced rosemary and 1/3T of salt and 1t of black pepper in a mixing bowl.
5. Remove bream to a baking bowl lined with method (4) evenly, then sprinkle with bread crumbs. Bake in oven at 230C on upper and lower element for about 5 minutes until done. Remove to serving plate.
6. Heat pan on medium, add 1 1/2T olive oil until smoking, stir-fry minced garlic and onion until through. Add shrimp and squid, sprinkle with 1/3t of black pepper and salt. Mix well and drizzle with white wine. Cook for about 30 seconds, then add diced tomato and soup broth, continue cooking for a minute, then stir in minced basil and remove. Drizzle over fish and serve.

Tips for a Perfect Dish

1. Taiwanese basil is also a variety of basil. Taiwanese basil darkens easily, but its aroma is stronger.
2. If dried herbs are used, use only half the portion of the fresh ones. Fresh rosemary is recommended in this recipe.

最想學會的外國菜

Part3 肉類

Meat Section

大快朵頤就是大塊吃肉，
大塊吃肉就是大快人心。
品味世界各地的風味料理，
試著在家中烹飪給親朋好友吃，
還有什麼比口腹之慾更容易滿足呢？
如果心中還有什麼忘不了的難過，
就吃吧！吃吧！

Eating the meat as much as you like is the real life.
To taste the whole world cuisines,
To cook at home for your lovely family and friends.
It's the most easy way to satisfy your appetite.
If you still have something worried or something sad,
Just eat it! Eat it mightily!

若以馬來和印尼風味醃料來比較，國人比較喜愛印尼風味，

因為馬來風味較甜，椰漿味過重，

而這裡CC老師帶來的配方足以溫暖全家人的胃。

The public might prefer Indonesian satay because Malaysian satay is sweeter with a strong coconut flavor. The recipe that Teacher CC offers here will satisfy the whole family's appetite.

沙嗲雞肉串
Chicken Satay

沙嗲雞肉串

材 料
去皮去骨雞腿肉（或雞胸肉）600 g.、竹籤數支

醃 料
無糖花生醬2匙、醬油2 1/2匙、黑胡椒粉1/2小匙、咖哩粉1小匙、蒜末2匙、白砂糖1小匙、薑泥1小匙、辣豆瓣醬1/2匙、洋蔥泥1匙

做 法
1. 雞肉切4×4公分寬，醃料拌勻，將雞塊放入醃漬約1晚。
2. 將雞肉一一串,入竹籤中，每支約串3～4塊。
3. 將雞肉串放進上下火230℃烤箱烤約8分鐘至熟（使用雙壁鍋、休閒鍋或材質好的平底鍋、不沾鍋：中火熱鍋，倒入2匙橄欖油，油熱後放入雞肉串煎熟即成）。

完美烹調寶典
Tips for a Perfect Dish
1. 醃漬雞肉若趕時間可將醃料材料加重1/2倍，醃漬2小時。
2. 雞肉若改成豬小里肌或牛肉，也別有一番口感。

Ingredients
600g chicken leg (boned and skin removed), bamboo skewers as needed

Marinade
2T unsweetened peanut butter, 2 1/2T soy sauce, 1/2t black pepper, 1t curry powder, 2T minced garlic, 1t white granulated sugar, 1t grated ginger, 1/2T hot chili bean paste, 1T grated onion

Methods
1. Cut chicken into 4cm x 4cm wide pieces. Combine ingredients from Marinade well and soak chicken in marinade overnight.
2. Skew chicken pieces with skowers, each skewer takes about 3-4 pieces.
3. Bake chicken skewer in oven at 230°C on the upper and lower element for about 8 minutes until done (use a Durotherm,hot pan or good quality frying pan, non-stick pan: Heat pan over medium and add 2T of olive oil, heat until smoking, fry chicken skewers in oil until done and remove).

Tips for a Perfect Dish
1. If you are in a hurry, double the ingredients in the marinade and marinate for 2 hours.
2. Chicken can be substituted with pork tenderloin or beef, which gives another kind of texture.

泰式椒麻雞
Thai Style Hot & Spicy Chicken

材 料 去骨雞腿2隻、蛋1個(打散成蛋液)、地瓜
粉1/4杯、大黃瓜1/4條(切片)、鹽1/2匙、
香菜(將葉子與莖分開,莖部切末1匙、
葉2匙)、紅辣椒末1匙、白醋2匙、蒜末1
匙、白砂糖1/2匙

醃 料 蒜末1匙、香菜莖末1匙、魚露1/3匙、醬
油11/2匙、白砂糖1/2匙、檸檬葉末1匙

椒麻汁 魚露1/2匙、醬油2匙、白砂糖11/2匙、
蒜末3匙、花椒粉1小匙、辣椒粉適量、
香油1匙、檸檬汁2匙

做 法

1. 大黃瓜和鹽拌勻,放置10分鐘後瀝乾水
   分;加入蒜末、白醋、白砂糖拌勻放置
   15分鐘,使其入味。

2. 將雞腿放入拌勻的醃料中浸泡約30分
   鐘,沾上蛋液、再均勻沾滿地瓜粉,放
   入160℃油鍋炸至呈金黃色,取出瀝乾
   油、切塊。

3. 盤中先放上醃漬入味的大黃瓜,放上雞
   塊,椒麻汁材料拌勻淋上,撒點香菜和
   紅辣椒末即成。

日式南蠻炸雞
Japanese Nanban Chicken Leg

材 料 去骨雞腿3隻(切塊)

醃 料 淡色醬油3匙、香蒜粉1小匙、白胡椒
粉少許、鹽少許

麵 衣 低筋麵粉3/4杯、蛋黃3個、冰水1 1/4
杯、橄欖油1匙

沾 醬 美乃滋1條(100g.)、水煮蛋1個(切細
末)、洋蔥末2匙(切極細)、檸檬汁1
匙、巴西里1/3小匙

做 法

1. 醃料拌勻，放入雞肉醃約1小時。
2. 拌勻麵衣材料，均勻覆蓋在雞肉上，放入
   160℃油鍋炸至呈金黃色，取出瀝乾油，
   放置盤中。上桌時附上一小碟沾醬。

Ingredients
3 chicken legs (boned and cut into pieces)

Marinade
3T light soy sauce, 1t flavored garlic powder, white
pepper and salt as needed

Coating
3/4C cake flour, 3 egg yolks, 1 1/4C ice water, 1T olive
oil

Nanban Dippping Sauce
1 pack mayonnaise (100g), 1 hard-boiled egg (minced),
2T minced onion (extremely fine), 1T lemon juice, 1/3t
parsley

Methods
1. Combine the ingredients for the marinade and marinate
   the chicken pieces for about 1 hour.
2. Combine the ingredients for the coating and spread
   evenly over the chicken. Deep-fry in oil at 160℃ until
   golden brown. Remove and drain, then place in serving
   plate. Serve with dippping sauce on the side.

Ingredients
2 chicken legs (boned), 1 egg (beaten), 1/4C yam flour,
1/4 cucumber (sliced), 1/2T salt, cilantro (separate leaves
from stems, 1T minced stems and 2T leaves), 1T minced
red chili pepper, 2T white vinegar, 1T minced garlic, 1/2T
white granulated sugar

Marinade
1T minced garlic, 1T minced cilantro stems, 1/3T fish
sauce, 1 1/2T soy sauce, 1/2T white granulated sugar,
1T minced lemon grass

Hot & Spicy Sauce
1/2T fish saue, 2T soy sauce, 1 1/2T white granulated
sugar, 3T minced garlic, 1t Szechwan peppercorn powder,
chili powder as needed, 1T sesame oil, 2 2T lemon juice

Methods
1. Marinate cucumber slices with salt and sit for 10 minutes,
   then drain well. Add minced garlic, white vinegar and
   white granulated sugar to mix. Let sit for 15 minutes until
   the flavor is absorbed.
2. Combine the ingredients from the marinade well, soak
   chicken legs in the marinade for about 30 minutes, then
   remove and coat with egg first, then with yam flour evenly.
   Deep-fry in oil at 160℃ until golden and remove, then cut
   into pieces.
3. Line the serving plate with cucumber slices, then top with
   chicken pieces. Drizzle hot and spicy sauce over on top,
   sprinkle with minced cilantro and red chili pepper. Ready
   to serve.

這道澳門名菜,帶著咖哩香及椰奶香,
是喜愛咖哩味的人,一定要學會的知名料理。

With a heavy curry and coconut flavor, this is Macao's famous dish.
Whoever loves curry should learn this famous dish.

葡國雞
Portuguese Chicken

葡國雞

材料

帶骨雞腿3隻、馬鈴薯1個(切塊)、沙拉筍1個(切塊)、蒜末1匙、紅蔥頭末1匙、洋蔥丁1/2個、椰奶1/2罐、奶水1/2罐、吉士粉1/2匙、麵粉3匙(過篩)、奶油2匙、咖哩粉3匙、高湯1/3杯(一般鍋1杯)

醃料

咖哩粉2匙、蒜末1匙、鹽1/2匙

做法

1. 醃料拌勻。雞腿剁成4塊,均勻抹上醃料放置約30分鐘。
2. 將雞塊放入160℃油鍋,以中火炸至呈金黃色取出瀝乾油;放入馬鈴薯炸至上色(或放入不沾鍋煎至呈金黃色,或入烤箱上下火230℃烤至呈金黃色)。
3. **製作葡汁**:中火熱鍋,倒入奶油和1 1/2匙橄欖油,煮至奶油融化,加入蒜末和紅蔥頭末爆香,轉小火慢慢倒入麵粉炒香,再放入咖哩粉和占士粉,炒至香味溢出,注入椰奶、奶水和高湯,攪拌均勻。
4. 放入雞塊和沙拉筍煮約10分鐘(雙壁鍋或休閒鍋只要5分鐘),再加馬鈴薯塊煮約5分鐘(雙壁鍋或休閒鍋只要3分鐘),以鹽調味即成。

完美烹調寶典
Tips for a Perfect Dish

1. 可改用全雞或半隻雞。
2. 葡汁亦可用來焗豆腐或做蔬菜類料理。
3. 吉士粉又稱為卡士達粉、雞蛋粉,可當成炸雞麵糊的材料之一,通常在烘焙材料行可買到。

Ingredients

3 chicken legs with bones, 1 potato (cut into pieces), 1 bamboo shoot (cut into pieces), 1T minced garlic, 1T minced shallots, 1/2T diced onion, 1/2 can coconut milk, 1/2 can evaporated milk, 1/2T cheese powder, 3T flour (sifted), 2T butter, 3T curry powder, 1/3C soup broth (ordinary measuring cup)

Marinade

2T curry powder, 1T minced garlic, 1/2T salt

Methods

1. Combine ingredients from marinade well. Chop each chicken leg into four pieces, then coat evenly with marinade and let sit for 30 minutes until flavor is absorbed.
2. Deep-fry chicken in 160℃ oil over medium heat until golden and remove. Deep-fry potato until brown (or fry in non-stick pan until golden, or bake in oven at 230℃ on upper and lower element until golden).
3. To prepare the Portuguese sauce: Heat pan on medium, add butter and 1 1/2T of olive oil. Cook until butter melt, add minced garlic and shallots, stir until fragrant. Reduce heat to low and fold in flour gradually. Stir gently until the fragrance is released, season with curry powder and cheese powder. Stir constantly until flavor is developed, pour in coconut milk, evaporated milk and soup broth. Stir until evenly-mixed.
4. Return chicken and potato, cook for about 10 minutes (it only takes 5 minutes to cook in a Durotherm or hot pan). Add potato chunks and cook for 5 minutes longer (3 minutes for a Durotherm or hot pan). Season with salt to taste. Serve.

Tips for a Perfect Dish

1. Chicken legs can be substituted with a whole chicken or half a chicken.
2. Portuguese sauce can be used in cooking tofu or vegetable dishes.
3. Cheese powder is also known as custard powder or powdered egg. It is one of ingredients in preparing deep-frying batter. It can be found at baking supply stores.

CC將馬來菜改變了一些，讓一般大眾更能接受馬來料理。

其實馬來菜還是有受中國飲食的影響，

而這道料理是CC最愛的馬來菜，與大家分享。

Teacher CC changes Malaysian dishes a bit to let the public know about them. In fact, Malaysian cuisine has been influenced by Chinese. This is CC's favorite Malaysian dish, introduced for you.

馬來風味雞
Malaysian Flavor Chicken

馬來風味雞

材料

去骨雞腿3隻、蒜末1 1/2匙、香菜末1 1/2匙、咖哩葉1小匙、咖哩粉1/3匙

醃 料

醬油2匙、白砂糖1小匙、蒜末1匙

醬 料

梅子醬2匙、海山醬1匙、高湯5～6匙(雙壁鍋或休閒鍋3匙)、醬油2匙、太白粉1匙

做 法

1. 醃料拌勻。醬汁材料拌勻備用。雞腿切塊，均勻抹上醃料放置約30分鐘。
2. 將雞塊放入160℃油鍋，以中火炸至呈金黃色，取出瀝乾油。
3. 另取一鍋，中火熱鍋，放入1匙油，加入蒜末、香菜末炒香，轉小火倒入咖哩粉、咖哩葉炒至香味溢出；倒入炸過的雞塊，轉中火並放入醬汁炒勻即成。

完美烹調寶典
Tips for a Perfect Dish

1. 如果不喜歡油炸可改以下列兩種方式處理雞：A.放3匙油於不沾鍋內煎至呈金黃色。B.放入上下火230℃烤箱烤約15分鐘至呈金黃色（時間視各家廠牌而略有出入）。
2. 如沒有梅子醬可選用自家醃的紫蘇梅，將籽取出，梅肉剁碎，加一些醃梅汁拌勻即可。
3. 咖哩葉散發出咖哩香味，尤其將其搗碎時更加明顯，常用於印度咖哩中。

Ingredients

3 boneless chicken legs, 1 1/2T minced garlic, 1 1/2T minced cilantro, 1t curry leaves, 1/3T curry powder

Marinade

2T soy sauce, 1t white granulated sugar, 1T minced garlic

Sauce

2T plum sauce, 1T hoisen sauce, 5~6T soup broth (3T for a Durotherm or hot pan), 2T soy sauce, 1T cornstarch

Methods

1. Combine the ingredients to make the marinade and the sauce. Cut chicken legs into small pieces and let sit in marinade for about 30 minutes.
2. Deep-fry chicken in 160℃ oil over medium heat until golden and remove to drain.
3. Heat another pan on medium, add 1T of cooking oil until smoking, add minced garlic and cilantro and heat until fragrant. Reduce heat to low and stir-fry curry powder and curry leaves until flavor is released. Return chicken and turn heat up to medium, then add the sauce and mix well.

Tips for a Perfect Dish

1. If deep-frying is not desired, there are two ways to handle the chicken: A. Add 3T of oil to a non-stick frying pan and fry chicken until golden. B. Bake in oven at 230℃ on the upper and lower element for about 15 minutes until golden (cooking times will differ by type of oven).
2. If plum sauce is not available, use home-made pickled plums. Remove and discard seeds, chop the plum finely and mix well with a little pickle juice.
3. Curry leaves are filled with curry flavor, especially when they are ground. It is a very common item in Indian cuisine.

這道料理迷倒很多人，連外國朋友也超級喜愛，直說過癮！

This cuisine charms everyone. Even my foreign friends love it
and think it is terrific and enjoyable!

墨西哥風味烤雞腿
Mexican Roast Chicken

深受好評！一定要試試看！
CC大受歡迎的不放油不洗鍋料理

烹煮各國料理，只要有一只好鍋，就能省略很多煩人的洗鍋工夫，還有烹調的煩瑣次序。這裡先來教3道料理，墨西哥風味烤雞腿+加州風味培根玉米（P.24）+泰式海鮮涼拌（P.21）一氣呵成，是宴客時省時省力的最佳選擇。

1. 直接用「墨西哥風味烤雞腿」的鍋，以烤雞的餘油做「加州風味培根玉米」，不必洗鍋也不再放油了。
2. 接著用「加州風味培根玉米」的鍋，製作「泰式海鮮涼拌」，將海鮮直接放入培根玉米的鍋內蒸煮3分鐘就OK了。不必洗鍋也不必放水和油喔！
PS.一般鍋因鋼材之故，較不適合作不洗鍋料理。

墨西哥風味烤雞腿

材料

去骨雞腿3隻、巧達起司100g.(切片)、香菜末1 1/2匙、黑胡椒和鹽適量

墨西哥莎莎醬

牛蕃茄1個(切小丁)、洋蔥末2匙、蒜末1匙、檸檬汁1/2匙、橄欖油1/2匙、鹽1/2匙、黑胡椒1/2小匙、巴西里少許

做法

1. 雞腿正反面撒上黑胡椒鹽(比例2：1)，有吃辣者可撒上適量的紅辣椒粉。
2. 將墨西哥莎莎醬材料拌勻，喜愛辣的人可再加少許紅辣椒末或紅辣椒粉。
3. 中火熱鍋，倒入1 1/2匙橄欖油，油熱後(油紋產生)放入雞腿，皮朝下煎至皮呈金黃色，起鍋移入烤箱上下火230℃烤約15～20分鐘(視各廠牌烤箱時間)，取出塗上墨西哥莎莎醬、放上起司，烤至起司融化稍呈金黃色即可取出，盛盤時撒上香菜末。
4. 如使用雙壁鍋或休閒鍋：中火熱鍋，不需放油，待鍋熱將雞皮朝下放入鍋內，轉中小火煎約2分鐘30秒，此時會釋放出油脂，將雞腿翻面，皮已呈金黃色，皮朝上淋上墨西哥莎莎醬蓋上鍋蓋改小火烤約5分鐘，打開鍋蓋鋪上起司，再烤約1～2分鐘即可取出置盤內，撒上香菜。

完美烹調寶典
Tips for a Perfect Dish

做好的墨西哥莎莎醬也可塗在法國麵包上，或搭配墨西哥脆片。

Ingredients

3 boneless chicken legs, 100g cheddar cheese (sliced), 1 1/2T minced cilantro, black pepper and salt as needed

Mexican Salsa Sauce

1 beefsteak tomato (diced finely), 2T minced onion, 1T minced garlic, 1/2T lemon juice, 1/2T olive oil, 1/2T salt, 1/2t black pepper, parsley as needed

Methods

1. Sprinkle black pepper salt (2:1) on both sides of chicken legs. Red chili powder can be added if desired.
2. Combine the Salsa sauce ingredients together, add minced red chili pepper or chili powder if desired.
3. Heat pan over medium, add 1 1/2T of olive oil until smoking, fry chicken legs with the skin facing down until golden. Place in oven at 230℃ on upper and lower element for about 15 to 20 minutes (depending on the brand of the oven), then remove and brush with salsa sauce, then top with cheese. Return to oven and bake until the cheese melts and becomes golden. Remove to a serving plate and sprinkle with minced cilantro.
4. Use a Durotherm or hot pan: Heat pan on medium (no oil is needed), wait until the pan is hot, place chicken legs with skin facing down in pan. Reduce to medium low heat and fry for 2 minutes and 30 seconds until chicken oil released. Turn the golden chicken legs over, with the skin facing up, drizzle salsa sauce over. Cover and reduce heat to low and fry for about 5 minutes. Remove top and spread cheese on top, then fry for about 1 to 2 minutes until done. Remove to a serving plate and sprinkle with cilantro. Serve.

Tips for a Perfect Dish

· Salsa sauce can be spread over French bread, or served with Mexican taco chips.

Highly praised! Must try! CC's popular no-oil no-cleaning afterwards cuisine!

When preparing foreign dishes with one good pan, many troublesome and complicated cleaning and cooking procedures can be skipped. Here come three dishes: Mexican Roasted Chicken + California Bacon Corn + Thai Style Seafood Salad, all at once. This is the best way to save time and energy when you are having a party.

1. Use the remaining oil from the pan for the Mexican Roast Chicken to cook the California Corn with Bacon. No need to rinse the pan and no oil is added.
2. Use the pan in which the California Corn with Bacon was cooked to prepare the Thai Style Seafood Salad. Just cook seafood directly in the pan for 3 minutes. No need to clean the pan and no water and oil is needed!
PS. In China, pans are sometimes not washed on the outside after cooking. This is not appropriate if you are using an ordinary pan because the steel is unsuitable for this traditional practice.

香濃的地中海風味，
會使你不知不覺中多吃了好幾片法國麵包。

This delicious mediterrean dish is so good that it makes
you eat extra French bread before you know it!

地中海菇蕈雞
Mediterrean Mushroom Chicken

地中海菇蕈雞

材料

雞1隻(去頭腳)、培根5片(切約1公分小丁)、洋蔥1個(切丁)、蒜末2匙、蘑菇1罐(去水)、新鮮香菇200g.(切片)、牛肝菌80g.、白酒11/2杯、雞高湯1,200～1,600c.c.(雙壁鍋或休閒鍋：800c.c.)、橄欖油2匙、黑胡椒和鹽適量

做法

1. 雞1隻剁成8塊(可請雞販幫忙剁)，洗淨擦乾水份，均勻抹上黑胡椒和鹽。

2. 牛肝菌泡溫水15分鐘後取出，水不用倒掉留著備用。

3. 中火熱鍋，倒入2匙橄欖油，油熱後(油紋產生)放入雞塊煎至呈金黃色取出，隨即放入培根煎至焦黃，倒入蒜末、洋蔥炒透。再加入煎過的雞塊，倒入白酒，待蒸發後放入高湯煮約30分鐘（使用雙壁鍋或HOT PAN鍋：只要18分鐘）。

4. 另取一鍋放入1匙油，倒入牛肝菌炒香，加入蘑菇和香菇炒香，倒入做法3.鍋中，再倒入浸泡牛肝菌的水，續煮12分鐘（使用雙壁鍋或休閒鍋：只要6分鐘），最後以黑胡椒和鹽調味。

完美烹調寶典
Tips for a Perfect Dish

1. 黑胡椒和鹽以2：1拌勻，就是完美比例不會太鹹或太淡。

2. 洋蔥一定要炒透，及炒至呈淡茶褐色，湯頭才會甘醇。

3. CC選用雙壁鍋來烹調此道料理，可省略1倍的烹調時間，且高湯量也可省了近1倍的量，因雙壁鍋鎖水性較強的關係。

Ingredients

1 chicken (head and legs removed), 5 slices bacon (diced into cubes approximately 1cm wide), 1 onion (diced), 2T minced garlic, 1 can mushroom (water removed), 200g fresh shiitake mushrooms (sliced), 80g Boletus mushrooms, 1 1/2C white wine, 1200~1600c.c. chicken soup broth (800c.c. if Durotherm or hot pan is used), 2T olive oil, black pepper and salt as needed

Methods

1. Chop chicken into 8 large pieces (have the vendor chop it), rinse and dry with paper towel, then coat evenly with black pepper and salt.

2. Soak Boletus mushrooms in warm water for 15 minutes, remove and reserve the water for later use.

3. Heat pan on medium, add 2T of olive oil and heat until smoking, fry chicken pieces until golden brown and remove. Use the remaining oil to fry bacon slices until brown, add minced garlic and onion. Stir-fry until done, return chicken pieces and add white wine. Cook until the wine is evaporated, pour in soup broth and cook for about 30 minutes (Use a Durotherm or hot pan: cook for 18 minutes).

4. Heat 1T of oil in another pan and stir Boletus mushrooms until fragrant, add mushrooms and shiitake mushrooms. Stir until the flavor is released, pour into method (3) along with the water for soaking Boletus mushrooms. Continue cooking for 12 minutes longer (use a Durotherm or hot pan: cook for 6 minutes only). Season with black pepper and salt to taste. Remove and serve.

Tips for a Perfect Dish

1. Black pepper and salt (2:1) is the best combination, not too salty or light.

2. Onion has to be fried until transparent and light brown, so that the broth is sweet and good.

3. CC uses a Durotherm to prepare this dish to save half of the cooking time, and the amount of soup broth is halved because the Durotherm seals in the liquid effectively.

我的學生告訴CC，他學會這道料理後已經烤了近150隻雞
給家人好友和自己吃，可想而知它受歡迎的程度有多高啊！

CC's student told her that since she learned this dish, she had cooked 150 chickens for her friends and family. You can imagine how popular this dish is!

繽紛烤全雞
Roasted Chicken with Veggie

繽紛烤全雞

材 料

全雞1隻、馬鈴薯2個(切滾刀塊)、南瓜200g.(切滾刀塊)、綠花椰菜150g.(切小朵)、紅黃甜椒各1個(斜切菱角型)、大蒜12粒、洋蔥1/2個(切塊)、月桂葉1片

醃 料

鹽2匙、黑胡椒4匙、迷迭香末3匙、鼠尾草末2匙、蒜末2匙、香蒜粉1/2匙、百里香末2匙、紅椒粉1匙(不吃辣改使用匈牙利紅椒粉)、檸檬汁1個

做 法

1. 拌勻醃料，留1匙備用(塞入雞腹)。雞洗淨拭乾水份，抹上醃料放置1晚(如急用至少需放置2小時，材料需加倍)，烤前將肚內塞入大蒜5粒、洋蔥塊50g.、月桂葉及醃料1匙。

2. 將全雞放入烤箱上下火220℃，烤約1小時至1小時半(視各廠牌烤箱而略有出入)，待雞肉快熟時放入蔬菜，需不時淋上烤出的雞油，才不會使蔬菜太乾而無光澤。

3. 使用休閒鍋：中火熱鍋，倒入2匙油，油熱後(油紋產生)放入醃好的雞煎至呈金黃色，放入剩下的大蒜和洋蔥，蓋上鍋蓋轉小火烤約25分鐘(約6～7分鐘時要翻面)，放入馬鈴薯塊蓋上鍋蓋續煮約3分鐘，再放入南瓜煮約4分鐘，待南瓜熟了再倒入綠花椰菜、甜椒，蓋上鍋蓋煮至冒煙即可熄火，放入外鍋直接端上餐桌食用。

完美烹調寶典
Tips for a Perfect Dish

1. 香草可使用乾燥的，但量需少1倍，建議至少迷迭香使用新鮮的。迷迭香較易栽種。

2. 使用休閒鍋烤的雞肉較不柴，也較有湯汁。

Ingredients

1 whole chicken, 2 potatoes (cut into irregular chunks), 200g pumpkin (cut into irregular chunks), 150g broccoli (cut into small florets), 1 each red and yellow bell pepper (cut into diamond shape pieces), 12 cloves garlic, 1/2 onion (cut into pieces), 1 bay leaf

Marinade

2T salt, 4T black pepper, 3T minced rosemary, 2T sage. 2T minced garlic, 1/2T flavored garlic powder, 2T minced thyme, 1T red chili powder (if spiciness is not desired use Hungarian paprika), 1 lemon (squeezed into juice)

Methods

1. Combine the ingredients for the marinade, retain 1 tablespoon for stuffing. Rinse chicken and dry. Coat evenly with marinade and sit overnight (in an emergency, let sit at least 2 hours). Stuff the chicken with 5 cloves of garli, 50g of onion pieces, bay leaf and 1 tablespoon of marinade.

2. Roast in oven at 220℃ on the upper and lower element for 60-90 mins (depending on the type of oven, the baking time may differ) until almost done. Add vegetables and drizzle with chicken oil released from the chicken to prevent the vegetables from becoming too dry and dull looking.

3. Use a hot pan: Heat pan over medium, add 2T of cooking oil until smoking, fry chicken until godlen. Add remaining garlic and onion. Cover and reduce heat to low, cook for about 25 minutes (turn chicken around every 6-7 minutes), then add potato and cover. Continue cooking for about 3 minutes longer, then add pumpkin and cook for 4 minutes until done. Add broccoli and bell peppers. Cover and cook until steam is released. Remove from heat and serve directly to the table.

Tips for a Perfect Dish

1. Reduce the ingredients by half if dried herb is used. Fresh rosemary, easier to plant, is strongly recommended.

2. The texture of the chicken is juicier and softer.

這道料理大宴小酌皆宜，宴客、拜拜、年節菜、居家解饞統統適合，
做法簡單又具五星級賣相，CC老師常常把鴨滷好放涼、
以錫箔紙鋪在市價不到100元的竹籃內，再用玻璃紙包裹鴨，
以緞帶綁個蝴蝶結包裝，當作伴手禮送給好朋友，相當受歡迎！

This dish is good for every occasion, feasting, worshiping the gods, or New Year's. The methods are simple and the resultant appearance is appealing. Teacher CC always gives it away as a handy present by cooling the stewed duck, lining aluminum foil at the bottom of bamboo basket worth less than 100 dollars, then wrapping the duck up with a piece of cellophane with a ribbon to garnish. Everyone loves it.

東南亞美味滷鴨
South East Asian Stewed Duck

東南亞美味滷鴨

材料

鴨1隻或光鴨(只沒有鴨頭、鴨翅、鴨腳的鴨,但也較便宜)、萵苣生菜150g.

A料:蒜粒1/2杯、南薑6片、八角3粒、小茴香1/2小匙、香菜頭2個

滷汁

醬油3/4杯、老抽1/4杯、魚露3大匙(參考量)、白砂糖(或冰糖)4匙、水3杯

沾醬

檸檬汁3匙、魚露11/2匙、白砂糖3匙、蒜末、香菜末各11/2匙、紅辣椒1匙

做法

1. 鴨洗淨瀝乾水分。
2. 鍋中放入滷汁料的醬油、老抽、魚露、和糖,煮滾後放入鴨,煮至上色後倒入水和材料A料滷約60分鐘(雙壁鍋或休閒鍋只需約30分鐘),放涼後剁塊。
3. 將萵苣生菜鋪在盤邊,中間放上鴨肉塊,淋上滷鴨汁。沾醬拌勻後放小盤中一起盛盤。

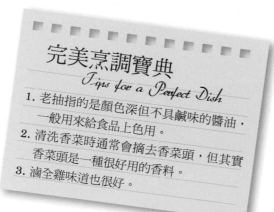

完美烹調寶典
Tips for a Perfect Dish

1. 老抽指的是顏色深但不具鹹味的醬油,一般用來給食品上色用。
2. 清洗香菜時通常會摘去香菜頭,但其實香菜頭是一種很好用的香料。
3. 滷全雞味道也很好。

Ingredients

1 duck or duck without head, wings and feet (it is cheaper), 150g iceberg lettuce

A Ingredients

1/2C garlic cloves, 6 slices tumeric, 3 star anises, 1/2t cumins, 2 cilantro roots

Stewing Sauce

3/4C soy sauce, 1/4C dark soy sauce, 3T fish sauce (see reference), 4T white granulated sugar (or rock sugar), 3C water

Dipping Sauce

3T lemon juice, 1 1/2T fish sauce, 3T white granulated sugar, 1 1/2T minced garlic, 1 1/2T minced cilantro, 1T red chili pepper

Methods

1. Rinse duck well and drain.
2. Heat soy sauce, dark soy sauce, fish sauce and sugar from stewing sauce until boiling, add duck and cook until brown. Add water and Ingredient A, stew for about 60 minutes (it takes only 30 minutes in a Durotherm or hot pan), remove and cool, then chop into serving size.
3. Line lettuce leaves on the side of the serving plate. Place duck in the center and drizzle with stewing sauce. Combine the ingredients from dipping sauce well and serve on the side as a dip.

Tips for a Perfect Dish

1. Dark soy sauce is only dark in color, it actually lacks a salty flavor. It is used for browning the ingredients.
2. The root of cilantro is usually discarded when we clean the cilantro. It is in fact a very useful herb for cooking.
3. Duck can be substituted with chicken, which also tastes wonderful.

越南香料五花肉
Vietnamese Herbal Pork

材 料 五花肉600g.、蔥末3大匙、黑胡椒適量

醃 料 紅蔥頭末2匙、蔥末3匙、香菜末3匙、紅辣椒末適
量、白砂糖1匙、魚露1匙、鹽1匙

沾 醬 檸檬汁1 1/2匙、蒜末1匙、紅辣椒1/2匙、魚露1/2
匙、白砂糖2匙、水3匙

做 法

1. 醃料拌勻，放入五花肉醃約2小時。沾醬材料拌勻備用。
2. 中火熱鍋，倒入2匙油，油熱後(油紋產生)放入五花肉煎
烤至熟，取出切片即可，上桌時附上沾醬食用。

完美烹調寶典
Tips for a Perfect Dish

1. 也可使用肋排或夾心肉、梅花
肉、小里肌，口感更有勁。
2. 沾醬也可做涼拌菜，檸檬汁可
隨個人喜愛的酸度做調整。

泰式茄醬肉末
Thai Style Pork with Tomato Sauce

材 料
豬絞肉600g.、蝦米1/3杯(切碎)、牛蕃茄4個(去皮切丁)、紅蔥頭末3大匙、蒜末3匙、紅咖哩醬2匙、香菜末2匙

調味料
魚露1/2匙、椰子糖(或白砂糖)1匙、泰國辣椒粉適量

做 法

1. 中火熱鍋，倒入2匙油，油熱後加入絞肉炒散，取出備用。

2. 紅蔥頭放入鍋內炒至金黃色，再倒入蒜末炒香，隨即放入蝦米炒香，再加入紅咖哩醬拌勻，倒入絞肉和蕃茄丁炒約6分鐘，最後放入調味料炒勻即成。盛盤時撒上香菜末。食用時搭配蔬菜或拌麵飯吃。

Ingredients
600g ground pork, 1/3C dried miniature shrimp (finely chopped), 4 beefsteak tomatoes (peel and diced), 3T minced shallots, 3T minced garlic, 2T red curry paste, 2T minced cilantro

Seasonings
1/2T fish sauce, 1T coconut sugar (white granulated sugar), Thai chili powder as needed

Methods
1. Heat pan over medium, add 2T of oil and heat until smoking, add ground pork and stir until white, then remove.
2. Stir-fry minced shallots until golden brown, add minced garlic and heat until fragrant, then add dried shrimp and stir until flavor is released, add red curry paste to mix. Return ground pork and add diced tomatoes. Saute for about 6 minutes and season with seasonings to taste. Mix well and remove to a serving plate. Sprinkle with minced cilantro and serve. You can also have vegetables, rice or noodles while you enjoy this.

Ingredients
600g belly pork, 3T minced scallion, black pepper as needed

Marinade
2T minced shallots, 3T minced scallion, 3T minced cilantro, minced red chili pepper as needed, 1T white granulated sugar, 1T fish sauce, 1T salt

Dipping Sauce
1 1/2T lemon juice, 1T minced garlic, 1/2T red chili pepper, 1/2T fish sauce, 2T white granulated sugar, 3T water

Methods
1. Combine ingredients for marinade well and marinate pork for 2 hours. Combine ingredients for dippping sauce together.
2. Heat pan over medium, add 2T of oil until smoking, fry pork until done on both sides. Remove and cut into slices. Serve with dippping sauce on the side as a dip.

Tips for a Perfect Dish
1. Pork ribs, shoulder meat, any fatty cut, or tenderloin can be used instead. The texture will be even chewier.
2. Dipping sauce can be used in cold dishes. Adjust the portion of the lemon as desired.

這是我在曼谷旅遊時，泰國人教CC的菜色，她特別買了打拋葉及泰國辣椒膏讓CC知道，
辣椒膏哪個牌子是她喜愛用的，且味道也是她認同的，而打拋葉與九層塔長得真像呢！

This dish is was taught to me by a Thai while I was travelling in Bangkok. She went through all the trouble to buy Thai Holy Basil leaves and Thai chili paste just to show CC. She picked the chili paste brand that she liked and the flavor she preferred. Thai Holy Basil leaves really look like basil!!

泰式打拋豬肉
Thai Holy Basil Pork

泰式打拋豬肉

材 料

豬絞肉400g.、打拋葉30g.、蒜末2匙、紅蔥頭末11/2匙、紅辣椒末3匙、泰國辣椒粉適量、泰國辣椒膏2匙、魚露1匙、椰子糖(棕櫚糖)1匙、高湯3匙(一般鍋1/4杯)

做 法

1. 中火熱鍋，倒入4匙油(材質好的平底鍋或不沾鍋只需約2匙油)，油熱後放入蒜末、紅蔥頭末爆香，加入紅辣椒末略炒幾下，放入絞肉炒散。放入泰國辣椒膏炒勻，倒入高湯、魚露、椰子糖和適量的泰國辣椒粉炒約8分鐘(材質好的平底鍋或不沾鍋只需約5分鐘)讓味道更入，最後放入打拋葉炒幾下即可盛盤。

完美烹調寶典
Tips for a Perfect Dish

1. 棕櫚糖是柬埔寨的特產，由棕櫚樹的果實提煉而成，味道近似黑糖，若無椰子糖(棕櫚糖)可使用白砂糖代替。
2. 有販售泰國材料的店也許會有新鮮的打拋葉，或打拋醬，若買不到也可以用以九層塔代替。
3. 豬絞肉以牛絞肉或雞肉片代替，別有一番風味。
4. 也可以加入小蕃茄(對切)一起炒。
5. 若在泰國的餐廳點這道料理，有的店家會再附上一個煎得香酥的荷包蛋。

Ingredients

400g ground pork, 30g Thai Holy Basil leaves, 2T minced garlic, 1 1/2T minced shallots, 3T minced red chili pepper, Thai style chili powder as eeded, 2T Thai chili paste, 1T fish sauce, 1T coconut sugar (palm sugar), 3T soup broth (ordinary measuring cup 1/4T)

Methods

1. Heat pan over medium, add 4T of oil until smoking (with a better quality frying pan or non-stick pan only 2T of oil is needed), add minced garlic and shallots and heat until fragrant. Add red chili pepper to mix, then saute ground pork until separated. Add Thai chili paste along with soup broth, fish sauce, coconut sugar and a little Thai style chili powder, stir-fry for about 8 mintues (it only takes 5 minutes with a better quality frying pan or non-stick pan) to ensure the flavor is well absorbed. Add Thai Holy Basil leaves and remove to a serving plate. Serve.

Tips for a Perfect Dish

1. Palm sugar is Cambodian specialty. It is refined from palm trees. The flavor is close to dark brown sugar. If coconut sugar (palm sugar) is not available, use white granulated sugar instead.
2. Thai grocery stores might carry fresh Thai Holy Basil leaves, or Thai Holy Basil paste; if not, use basil instead.
3. Ground beef or chicken slices may be used instead of pork in this dish.
4. Cherry tomato can be halved and added to the dish.
5. If you order this dish in a Thai restaurant, it might come with a fried egg as a side dish.

令人垂涎欲滴的美食，你絕不能錯過！是很有五星級感的主菜哦！
What a delicious gourmet dish! Not to be missed! It feels like the main dish
in a five star restaurant!

韓式松阪豬肉
Korean Matsuzaka Pork with Kimchi

肉類 Meat Section

韓式松阪豬肉

材 料

松阪豬肉4片600g.、泡菜150g.、韭菜100g.、醬油1小匙、香油1/2匙

醃 料

韓國辣椒醬2 1/2匙、黑胡椒粉1小匙、洋蔥泥1匙、蒜末1 1/2匙、香油1匙、醬油2匙

做 法

1. 醃料拌勻，放入豬肉醃約2小時。
2. 將剛洗淨的韭菜放入鍋內(不需放水)，蓋上鍋蓋以小火煮至冒煙韭菜即熟，此為無水烹調方式(一般鍋需放水，待水滾再放入韭菜汆燙熟)，熄火，取出瀝乾水份。
3. 取約1/5韭菜和70g.泡菜倒入調理機內，加醬油和香油，打成泥狀，鋪於盤中。
4. 中火熱鍋，倒入2匙油，待油熱放入醃好的豬肉煎熟，取出放入做法3.上。將剩餘的泡菜和韭菜放置盤邊搭配。

完美烹調寶典
Tips for a Perfect Dish

1. 此醃料亦可醃雞肉。
2. 可將肉切片，與醃料拌勻放置1小時後，做成韓國烤肉方式搭配生菜食用。

Ingredients

600g 4 slices neck pork , 150g Kimchi, 100g leek, 1t soy sauce, 1/2T sesame oil

Marinade

2 1/2T Korean chili paste, 1t black pepper, 1T grated onion, 1 1/2T minced garlic, 1T sesame oil, 2T soy sauce

Methods

1. Combine all the ingredients from marinade well, marinate the pork for about 2 hours.
2. Rinse leek, then cook in pan without water added over low heat until steaming and done. This is the so-called "waterless cooking" (an ordinary pan needs to be filled with water first, then brought to a boil before adding leeks). Remove from heat and drain.
3. Put 1/5 of leek and 70g of kimchi in a food processor, add soy sauce and seasame oil. Blend until thick and spread evenly over plate.
4. Heat pan over medium, add 2T of oil, heat until smoking, fry pork until done and remove to the top of method (3). Garnish the plate with the remaining kimchi and leeks. Serve.

Tips for a Perfect Dish

1. The marinade can be used to marinate chicken meat.
2. Pork can be sliced into small pieces and marinated for 1 hour, then barbecued and served with lettuce.

很多學生對中南美洲料理興趣缺缺，原因是口味不對、香料過重，
享用時無法產生幸福感。但這道菜好吃的程度
讓我的學生因此而對墨西哥菜改觀，你也來試試吧！

Many students are not very intersted in Central and South
American cuisine because the flavor is too heavy for
them and they cannot fully enjoy it. However, this dish is
so delicious that my students changed their minds about
Mexican cuisine. Why don't you try it too!?

墨西哥風味燉肉
Mexican Stewed Pork

墨西哥風味燉肉

材 料

豬絞肉(或牛絞肉)600g.、紅腰豆罐頭1罐、洋蔥末1杯、蒜末2匙、蕃茄粒罐頭1罐(搗碎)、巧達起司150g.、高湯11/2杯(一般鍋3杯)、巴西里少許、奧勒岡1小匙、小茴香粉1/2匙、紅辣椒粉適量

調味料

黑胡椒、鹽適量

做 法

1. 中火熱鍋,倒入2匙油,油熱後(油紋產生)放入絞肉炒散,先取出。加入蒜末炒香、隨即放入洋蔥末炒透,再放入絞肉拌炒均勻。

2. 放入搗碎的蕃茄粒,紅腰豆(需去除水)、高湯、奧勒岡1小匙、小茴香粉1/2匙、紅辣椒粉拌勻,蓋上鍋蓋,至冒煙後轉小火燉約1小時(雙壁鍋或休閒鍋只需約25分鐘);放入胡椒、鹽調味拌勻,最後鋪上起司、撒上巴西里即成。

完美烹調寶典
Tips for a Perfect Dish

1. 墨西哥料理中常用的起司是高達起司或巧達起司。

2. 絞肉要盡量炒散才不會有腥味,也才容易入味。

3. 以豆類而言,紅腰豆營養價值相當高,含有豐富的蛋白質、纖維、鐵質等,很適合素食者和孕婦。原產地為南美洲,台灣目前只有紅腰豆罐頭,在大型百貨公司的超市(如city super)有賣。

Ingredients

600g ground pork (or ground beef), 1 can red kidney beans, 1C minced onion, 2T minced garlic, 1 can whole tomatoes (crushed), 150g cheddar cheese, 1 1/2C soup broth (ordinary measuring cup 3C), parsley as needed, 1t oregano, 1/2T cumin powder, red chili powder as desired

Seasonings

salt and white pepper as needed

Methods

1. Heat pan over medium, add 2T of oil and heat until smoking, stir in ground pork and cook until white, remove. Add minced garlic and heat until fragrant, then minced onion until flavor is released, then return pork and mix well.

2. Add crushed tomatoes, red kidney beans (discard the water), soup broth, 1t oregano, 1/2T cumin powder and red chili powder to mix. Cover and cook until steam comes out, reduce heat to low and stew for about 1 hour (it takes only 25 minutes in a Durotherm or hot pan). Season with pepper and salt to taste. Spread the cheese on top and sprinkle with parsley. Ready to serve.

Tips for a Perfect Dish

1. The most common cheese in Mexican cuisine is gouda cheese or cheddar cheese.

2. Stir and separate the ground pork as much as possible to avoid a fishy taste, and enable the seasoning to be absorbed better.

3. As for beans, red kidney beans are very rich in nutrition, containing high protein, fiber, and iron. They are perfect for for vegetarians and pregnant women. Originally from South America, Taiwan carries only canned red kidney beans, which can be found in supermarkets in large department stores (e.g. City Super).

頂級藍帶豬排
Top Blue Belt Steak

這道料理雖然各家餐廳或老師都有推出，
但每個人做功不同，CC是被學生要求一定要學最經典最頂級口感的法式藍帶豬排
CC找了最佳組合起司和火腿，終於完美演出。

This top flavor dish is served in every restaurant and taught by all the cooking teachers. However, everyone has a different approach to it. CC was requested by the students to make the most classic, most tasty Blue Belt Pork Steak. CC put together the best cheese and ham -- here goes the perfect show!

頂級藍帶豬排

材料

小里肌肉600g.、火腿肉4片(約100g.)、富勒比起司4片(約100g.)、名店頂級土司4片、黑胡椒、鹽適量、萵苣生菜100g.(切絲)、蛋2個(打散)、麵粉1/4杯

沾醬

美乃滋100g.(1小條)、第戎醬1 1/2匙、檸檬汁1 1/2匙、巴西里1小匙

做法

1. 沾醬材料拌勻，放置10分鐘後再食用較美味。
2. 里肌肉切蝴蝶刀法，再用肉錘拍打使之鬆弛，撒上黑胡椒及鹽（比例2：1）。

蝴蝶刀法：
第一刀不切斷
第二刀切斷

3. 將起司及火腿鋪於肉的切面中間。
4. 製作麵包粉：土司放入烤箱，以上下火120℃烤約8分鐘至酥脆再刨成粗粉狀。
5. 醃好的里肌肉均勻沾上麵粉，再沾上蛋液、裹上麵包粉。放入160℃油鍋炸至呈金黃色，取出瀝乾油，切大片盛盤，裝飾生菜葉。

完美烹調寶典
Tips for a Perfect Dish

1. 小里肌選用黑毛豬才好吃。火腿我選用來自加拿大Freybe經典火腿，土司我選用FLAVOR FIELD的頂級土司(有台北101・台北SOGO復興館・高雄漢神巨蛋3家分店)，搭配富勒比起司，這是我的完美配方，吃過的都嘖嘖稱讚。
2. 富勒比乳酪(Fol Epi)產於法國 Pays de Loire，法文的原意是「野麥梗」，質地較結實，很適合直接品味或搭配紅酒，稍帶麥香、香醇甘美的味道會慢慢地從咀嚼中釋出。

Ingredients

600g pork tenderloin, 4 slices Freybe ham (approximately 100g), 4 slices Fol Epi cheese (approximately 100g), 4 pieces high classy bread (from top bakery), combination of black pepper and salt at 2:1, 100g lettuce (shredded), 2 eggs (scrambled), 1/4C flour

Dipping Sauce

100g mayonnaise (1 small pack), 1 1/2T Dijon mustard, 1 1/2T lemon juice, 1t parsley

Methods

1. Combine the ingredients from the dipping sauce, let sit for 10 minutes to ensure the flavor is well absorbed.
2. Cut pork tenderloin into a butterfly (one cut through and second cut half), hammer with a meat tenderizer and sprinkle with black pepper and salt to taste.
3. Stuff cheese and ham in the center of the cut.
4. To prepare coating for the meat: Bake bread in the oven at 120℃ on upper and lower element for about 8 minutes until crispy, then remove and grind finely.
5. Coat the tenderloin evenly with flour, then dip in eggs and coat evenly with coating. Deep-fry in oil at 160℃ until golden. Remove, drain well and cut into large pieces. Transfer to a serving plate and garnish with lettuce leaves. Serve.

Tips for a Perfect Dish

1. Select tenderloin from black pigs: it has better texture. I use classic Freybe ham from Canada and Flavor Field bread (fom Taipei 101, Sogo Fuhsin Branch Taipei, 3 Hanshin department store branches, Kaohsiung) to go with the Fol Epi cheese. This is my perfect combination of ingredients. Whoever tastes it always has nice things to say about it!
2. Fol Epi Cheese is from Pays de Loire France. It means "wild wheat stalk" in French. Its texture is firmer and it may be served directly or served with red wine. Its tasty wheat flavor can be slowly released while chewing.

這道層次口感的烤豬可做為主菜，也可將烤好的肉
切薄片做三明治、切小薄片做為前菜，是應用範圍很廣的特色料理。

This roasted pork with its many layers of texture can be used as a main dish. It can also be sliced and used in a sandwich, or as an appetizer. This special dish has a wide variety of uses.

加勒比海蒜烤豬
Caribbean Sea Garlic Roasted Pork

加勒比海蒜烤豬

材料
小里肌肉2條(約1,000g.)、大蒜1/2杯、洋蔥1個(切塊)、棉線2條、月桂葉1片、柳丁汁3匙、檸檬汁2匙、鹽2匙、紅辣椒粉適量

醃料
大蒜12顆、奧勒岡1/2匙、小茴香1/2小匙、洋蔥1/3個（切丁）、黑胡椒粒1匙

做法
1. 醃料放入調理機中打碎。
2. 小里肌肉塗抹上醃料和柳丁汁、檸檬汁、鹽、紅辣椒粉醃一晚，以棉線綁好。
3. 小里肌肉入鍋煎至呈金黃色取出，移入烤箱，鋪上蒜粒、洋蔥塊、月桂葉，以上下火230℃烤約25～30分鐘(視各家廠牌而略有出入)。
4. 使用休閒鍋：中火熱鍋，倒入2匙橄欖油，油熱後放入里肌肉煎至呈金黃，再放入去皮的蒜粒和洋蔥塊、月桂葉，蓋上鍋蓋以小火煮約15分鐘。取出切片即可盛盤。

完美烹調寶典
Tips for a Perfect Dish
1. 此料理方式使用牛肉或雞肉都很美味。
2. 綁上棉線可以固定肉體，以免煎烤時變形。

Ingredients
2 pork tenderloin (approximately 1000g), 1/2C garlic, 1 onion (sliced), 2 cotton strings, 1 bay leaf, 3T orange juice, 2T lemon juice, 2T salt, red chili powder as needed, 2T olive oil

Marinade
12 cloves garlic, 1/2T oregano, 1/2t cumin powder, 1/3 onion (diced), 1T black peppercorns

Methods
1. Combine the ingredients from the marinade, put in the food processor and blend until fine.
2. Brush the tenderloin with marinade, orange juice, lemon juice ,red chili powder and salt, let sit overnight and tie up with a piece of cotton string.
3. Fry tenderloin in pan until golden and remove to oven. Spread garlic cloves, onion pieces and bay lef over on top, then bake at 230℃ on upper and lower element for about 25 to 30 minutes (the time differs depending on the brand).
4. Use a hot pan: Heat pan on medium, add 2T of olive oil until smoking, fry tenderloin until golden, then add garlic cloves, onion, bay leaf, then cover and cook over low heat for about 15 minutes until done. Remove, slice and serve.

Tips for a Perfect Dish
1. Beef or chicken can be used in this delicious recipe.
2. Tie the meat with cotton string to secure its shape during cooking.

法式紅酒
蒜香燉牛肉
French Garlic Beef with Red Wine

好吃！好吃！真好吃！
學生們形容此道料理的感覺很直接哦！

"Yum, yum, really yummy!" This is what the students think about this dish!!

法式紅酒蒜香燉牛肉

材 料

牛腱1,000g.、蒜末3/4杯、奶油2匙、培根80g.、蕃茄粒罐頭1罐、紅酒11/2杯、檸檬皮1個、高湯1杯(一般鍋3杯)、黑胡椒和鹽適量

做 法

1. 牛腱一切為二，約6公分長，撒上黑胡椒、鹽醃約10分鐘。蕃茄粒放入調理機內打成泥狀備用。
2. 中火熱鍋，倒入1匙橄欖油，待油熱（油紋產生）放入牛肉，由鍋邊放進奶油，煎至兩面呈焦黃色取出。
3. 培根切寬約1公分，倒入做法2.鍋內，煎至有些焦黃，倒入蒜末和煎好的牛肉，淋上紅酒煮至蒸發(濃縮剩一半)。倒入蕃茄泥和高湯，改小火燉約2小時，取出牛肉放置盤中（使用壓力鍋，倒入蕃茄泥和高湯，蓋上壓力鍋鍋蓋，待上升二條紅線，改小火燉約18分鐘，即可取出牛肉）。
4. 將檸檬皮放進做法3.的湯汁內煮約3分鐘，以黑胡椒鹽調味後過濾，將過濾後的湯汁淋在牛肉上即成。

完美烹調寶典
Tips for a Perfect Dish

1. 紅酒一定要煮至蒸發，否則會有苦澀味。
2. 檸檬皮不要削到白色部位，會有苦澀味；如已削到白色部分可放入沸水內汆燙一下，即可除去苦澀味。
3. 如改用雞肉，酒要改成白葡萄酒，時間縮短，雞肉以土雞或仿土雞為佳。

Ingredients

1000g beef tendon, 3/4C minced garlic, 2T butter, 80g bacon, 1 can whole tomatoes, 1 1/2C red wine, 1 lemon peel, 1C soup broth (ordinary measuring cup 3C), combination of black pepper and salt at 2:1 as needed

Methods

1. Cut tendon into halves about 6cm long, sprinkle with black pepper and salt, then lend sit for 10 minutes. Blend tomatoes in a food processor until mashed.
2. Heat pan over medium, add 1T of olive oil until smoking, add beef and slide the butter down into the pan from the side. Fry the beef until light brown on both sides and remove.
3. Cut bacon into 1cm wide pieces and add to method (2). Fry until light brown, add minced garlic and return beef. Drizzle with red wine and cook until evaporated (condensed to half the amount). Pour in tomato and soup broth, reduce heat to low and stew for about 2 hours. Remove beef to a serving plate (use a Duromatic: pour tomato and soup broth in cooker, cover and cook until two red lines go up, reduce heat to low and stew for 18 minutes more. Remove beef to plate).
4. Cook lemon peel in liquid from method (3) for 3 minutes. Season with black pepper and salt to taste. Pour through a sieve to remove any impurities, and drizzle over beef. Serve.

Tips for a Perfect Dish

1. Red wine has to be cooked until evaporated, or the dish will taste bitter.
2. Do not use the white portion of the lemon skin, it is quite bitter. If you do, blanch in boiling water to remove the bitterness.
3. If chicken is used in this recipe, switch red wine to white wine and shorten the cooking time. Select free range chicken or similar chicken if needed.

普羅旺斯
燉牛腱
Provence Stewed Beef Tendon

有著濃濃歐式鄉村情調的特殊美味……

This dish is filled with a special thick country style flavor

普羅旺斯燉牛腱

材料

牛腱1,000g.、麵粉1/4杯、奶油2匙、牛高湯1,000c.c.、培根100g.(切約0.8公分小丁)、大蒜10粒(切片)、洋蔥2個(切絲)、紅蘿蔔丁1/2杯、西洋芹2支(切丁)、紅酒1杯，檸檬皮屑1個、柳丁皮屑1個、新鮮百里香5支(或乾燥1小匙)、月桂葉1片、黑胡椒粒1匙、丁香1/2匙、巴西里少許

做法

1. 牛腱切約5公分寬小塊，撒上黑胡椒和鹽（比例2：1），均勻沾上麵粉備用。
2. 奶油放入鍋內，以小火煮至融化，放入培根以中小火煎至酥脆取出；放入牛腱，利用鍋內餘油煎至呈金黃色，取出。
3. 倒入蒜片，再利用鍋中剩餘的油炒香；放入洋蔥絲炒透至呈淡菜褐色；放入紅蘿蔔丁、西芹丁，以小火慢炒3分鐘；再加入煎好的牛腱和培根，轉中大火，淋上紅酒煮約2分鐘（讓酒精蒸發）；倒入高湯、檸檬、柳丁皮屑和香料，燉約2個小時（使用壓力鍋：待壓力鍋上升二條紅線轉小火煮20分鐘即成）。
4. 取出牛腱盛盤，將湯汁過濾後淋上，撒些巴西里即成。

完美烹調寶典
Tips for a Perfect Dish

1. 洋蔥要炒透，甜味才會釋放出來，湯頭也會比較甘醇。
2. 酒一定要蒸發(1杯煮至1/2杯)，才不會有澀味，煮菜的紅酒不要選太甜的。
3. 刮檸檬皮屑、柳丁皮屑時，不要削到白色部分，否則會苦苦的。
4. 也可改用豬腱來做，但時間要縮短1.5倍。

Ingredients

1000g beef tendon, 1/4C flour, 2T butter, 1000c.c. beef soup broth, 100g bacon (cut into 0.8cm pieces), 10 cloves garlic (sliced), 2 onions (shredded), 1/2C diced carrot, 2 strings western celery (diced), 1C red wine, zest of one lemon, zest of one orange, 5 fresh thyme (or 1t dried thyme), 1 bay leaf, 11 black peppercorn, 1/2T cloves, parsley as needed

Methods

1. Cut beef tendon into 5cm pieces, sprinkle with black pepper and salt (2:1), then coat evenly with flour for later use.
2. Heat butter in pan over low heat until melted, sauté bacon on medium low until crispy and remove. Use the remaining oil to fry beef tendons until golden and remove.
3. Add garlic slices and stir fry until fragrant, add shredded onion and sauté until done and light brown, then add diced carrot and celery. Cook on low for about 3 minutes, then return beef tendon and bacon. Increase heat to medium high and drizzle with red wine, cook for about 2 minutes until the wine evaporated. Pour in soup broth, lemon zest and orange zest as well as all the spices, stew for about 2 hours (Use Duromatic, Cook until the two lines on the cover turn red, reduce heat to low and cook for 20 minutes)
4. Remove beef tendon to a serving plate, discard any impurities from the liquid and drizzle over beef, then sprinkle with parsley. Serve.

Tips for a Perfect Dish

1. The onions have to be sautéed until transparent and brown, so that the sweetness can be released and the soup will be sweet.
2. Wine has to be evaporated (from 1C to 1/2C) to prevent any dry flavor. Avoid any wine that is too sweet for cooking.
3. When peeling the lemon and orange for the zest, do not peel the white part, which is strongly bitter.

CC老師平常不愛吃羊肉或羊排，而唯一能接受的就是自己料理的香草羊排，
淡淡的香草味去除了羊騷，是CC老師和朋友們最愛吃的羊肉料理。

Teacher CC does not eat lamb or lamb chops. The only dish she can accept is
the herbal lamb chop she makes herself. The light herbal fragrance removes the
gamy flavor of the lamb. It is CC and her friends' favorite lamb dish.

義式香草羊排
Italian Herbal Lamb Chop

義式香草羊排

材 料

A. 羊小排4隻、黑胡椒和鹽適量
B. 迷迭香末2匙、百里香末2匙、鼠尾草末2匙、蒜末2匙、巴西里1/2匙

沾 醬

牛高湯3匙、蕃茄粒罐頭(搗碎連汁)3匙、巴薩米克醋11/2匙、奶油1/2匙

做 法

1. **製作沾醬：**巴薩米克醋放入鍋內以中火煮至剩約1/2量，倒入牛高湯、蕃茄粒及汁、黑胡椒和鹽少許，續煮至湯汁濃縮至一半，最後拌入奶油。
2. 羊小排均勻撒上黑胡椒和鹽(比例2：1)。再將材料B.拌勻抹上羊小排，放置約20分鐘。
3. 平底鍋中倒入11/2匙橄欖油，中火煮至油熱(產生油紋。)後放入羊排，先將有油脂的地方朝向鍋底煎至呈焦黃色，再平放煎熟兩面。
4. 煎好的羊排放盤中，淋上沾醬，可以新鮮迷迭香點綴。

完美烹調寶典
Tips for a Perfect Dish

1. 製作沾醬時若使用市售的高湯或湯塊，此時不要放鹽。
2. 煎羊排一定要先將有油脂的地方朝鍋底煎至焦黃色，才不會有羊騷味。
3. 這道料理的香草料中，迷迭香最好選擇新鮮的，口感會更好些。如果只喜歡迷迭香口味也可以只放一種香料。

Ingredients

A 4 lamb chops, black pepper and salt as needed
B 2T minced rosemary, 2T minced thyme, 2T minced sage, 2T minced, 1/2T parsley

Dipping Sauce

3T beef soup broth, 3T canned tomato (whole, crushed with its juice), 1 1/2T balsamic vinegar, 1/2T butter

Methods

1. To prepare dipping sauce: Cook vinegar in pan on medium until half of the amount is left. Pour in beef soup broth, whole tomatoes with its juice, black pepper and salt to season. Continue cooking until the sauce is condensed to half the amount. Stir in butter.
2. Sprinkle black pepper and salt evenly over lamb chops (2:1). Combine ingredient (B) well together and spread evenly over lamb chop. Marinate for about 20 minutes.
3. Add 1 1/2T of olive oil in frying pan, heat over medium until smoking, put in lamb chops with the part where more fat gathers facing down. Fry until yellow and burnt, then turn over and fry until done on both sides.
4. Remove the lamb chops to a serving plate, drizzle dipping sauce on top and garnish with fresh rosemary. Serve.

Tips for a Perfect Dish

1. If market-sold soup broth or flavor cubes are used in preparing dipping sauce, do not add salt to season.
2. Fry the side where more fat gathers when frying the lamb chop, to prevent the gamy flavor.
3. Fresh rosemary is strongly recommended in this herbal dish. The texture will be better. If you prefer rosemary only, you can skip the rest.

迷人的風味、令人為之著迷的好料理，
無論如何你一定要試試，一鍋上菜就可搞定一餐。

Charming flavor, fascinating dish. You must give it a try. One
pot on the table can solve a meal!

歐風燉什錦
European Farmer's Stew

完美烹調寶典
Tips for a Perfect Dish

1. 若使用新鮮的百里香，可在入鍋前以
   手拍三下，讓香味溢出再丟入鍋中。
2. 冬天可多放一些麵粉，讓煮出來的湯
   頭較濃稠；相反的夏天則要少放些麵
   粉，讓湯稀些，才不會感覺過熱！

歐風燉什錦

材 料

牛腱600g.、土雞半隻、馬鈴薯切滾刀塊2個(約200g.)、南瓜切滾刀塊(約200g.)、白綠花椰菜各150g.、牛蕃茄去皮切塊大的2個(或小的3個)、洋蔥末1/2杯、月桂葉1片、新鮮百里香4支(乾燥1小匙)、麵粉1/3杯、融化奶油2匙、牛高湯800 c.c.(雙壁鍋400c.c.)、雞高湯800 c.c. (雙壁鍋400c.c.)、橄欖油2匙、黑胡椒和鹽適量

做 法

1. 雞切塊，撒上黑胡椒和鹽（比例2：1），均勻沾上麵粉。牛腱切約2.5公分小塊，做法同雞塊。

2. 中火熱鍋，倒入2匙橄欖油，待油熱（油紋產生）放入雞塊煎至兩面呈金黃（先將雞皮那一面朝下），取出。牛腱的做法亦同，煎至金黃取出。

3. 奶油放入原鍋中，倒入洋蔥末以小火炒透（呈淡茶褐色），加入3匙麵粉炒勻，放入牛腱、牛高湯、雞高湯、牛蕃茄，以及月桂葉、百里香。

4. 蓋上鍋蓋，煮至冒煙後轉小火繼續煮約50分鐘，需不時攪拌最後放入煎好的雞塊，轉中火持續蓋住鍋蓋，煮至冒煙後再轉小火煮20分鐘。放入馬鈴薯塊煮一下，加入南瓜塊續煮，待馬鈴薯快熟時倒入花椰菜煮至熟熄火，以黑胡椒和鹽調味。

5. 使用雙壁鍋或休閒鍋：蓋上鍋蓋，煮至冒煙後轉小火繼續煮約25分鐘，放入煎好的雞塊，轉中火蓋住鍋蓋，煮至冒煙後再轉小火煮10分鐘。放入馬鈴薯塊煮一下，加入南瓜塊續煮，待馬鈴薯快熟時倒入花椰菜、蓋上鍋蓋，轉中火煮至冒煙隨即熄火，燜約15秒，最後以黑胡椒和鹽調味。此時的花椰菜是最好吃的脆度。

Ingredients

600g beef tendon, 1/2 free range chicken, 2 potatoes (approximately 200g, roll-cut into pieces), pumpkins (approximately 200g,roll-cut into pieces), 150g cauliflower, 150g broccoli, 2 beefsteak tomatoes (or 3 small tomatoes, skin removed, cut into large chunks), 1/2C minced onion, 1 bay leaf, 4 stalks fresh thyme (or 1t dried thyme), 1/3C flour, 2T melt butter, 800c.c. beef soup broth (400c.c. for a Durotherm), 800c.c. chicken soup broth (400c.c. for a Durotherm), 2T olive oil, black pepper and salt as needed

Methods

1. Cut chicken into pieces, sprinkle with black pepper and salt (2:1) to season, then coat evenly with flour. Cut beef tendon into pieces about 2.5cm wide. Preparation methods are the same as for the chicken.

2. Heat pan on medium, add 2T of olive oil and heat until smoking, fry chicken pieces until golden on both sides (fry the side with the skin on first), then remove. Fry beef tendon in the same oil until golden and remove.

3. Heat butter in pan, pour in minced onion, stir-fry over low heat until done through (light brown), add 3T of flour to mix. Return beef tendon, add beef soup broth, chicken soup broth, and beefsteak tomatoes as well as bay leaf and thyme.

4. Cover and cook until steaming, reduce heat to low and continue cooking for about 50 minutes. Stir constantly, then return chicken pieces. Increase heat to medium, with the top on, cook until steaming, then reduce to low again and cook for 20 minutes longer. Season with black pepper and salt to taste.

5. In Durotherm or hot pan: Cover and cook until steaming, reduce heat to low and continue cooking for about 15 minutes. Return fried chicken pieces, increase heat to medium and cook until steaming, then reduce heat to low and cook for 10 minutes more. Add potato pieces and pumpkin pieces and cook until half done, add cauliflower and broccoli. Cover and increase heat to medium, cook until steaming and remove from heat. Simmer for 15 seconds. Season with black pepper and salt. The cauliflower should be at its best crunchiness.

Tips for a Perfect Dish

1. If fresh thyme is used here, press it gently with hands three times to let the flavor released before adding to the pan.

2. Add a little more flour in the winter, so that the soup broth is thicker. Add less in the summer to make the soup thinner and clearer, so you won't feel as hot.

最想學會的外國菜
Part4 主食和飯類
Main Dish and Rice Sections

最想學會的外國菜
Part4 主食和飯類
Main Dish and Rice Sections

最想學會的外國菜
Part4 主食和飯類
Main Dish and Rice Sections

最想學會的外國菜
Part4 主食和飯類
Main Dish and Rice Sections

拿一只鍋、放一些米、加一些料，
細細的燉、慢慢的熬，
看著米粒由透明變白實，
看著料汁由清淡變稠濁，
嗅著香味，多麼有成就感的一碗米飯。
在冬天吃一碗扎扎實實的飯，
那種厚厚的感覺好幸福呀！

To find a pot、put some rice、put some sauce,
Cook them softly，steam them slowly.
The rice will change into white color
The sauce will change into thickened.
To feel the good smell,
You will find that it's a great bowl of rice.
What a great feeling of happiness
to enjoy such bowl of rice in the cold winter.

最 想 學 會 的 外 國 菜
Part4 主食和飯類
Main Dish and Rice Sections

泰國涼拌河粉
Thai Cold Mix Pho

清爽的涼拌口味可當主菜也可當下飯菜，
是CC在泰國時愛吃的料理之一。

This light clear, cold mixed dish can be used as main course
to help your appetite. This is CC's favorite dish in Thailand.

泰國涼拌河粉

材 料

泰國河粉1/2包、絞肉300g.、蝦米3匙、鳳梨1/4個(切片)、蒜末2匙、洋蔥末1/2杯、紅辣椒末2匙、香菜末6匙、芹菜末3匙、紅蔥頭3粒(切片)、碎花生1/4杯、泰國辣椒粉適量(依個人喜愛辣度)、鮮蝦200g.(去腸泥)、白砂糖適量、檸檬1個(切8瓣)

調味料
魚露2匙、檸檬汁1 1/2匙、酸子汁1匙、椰子糖1匙

做 法

1. 中火熱鍋，放入3 1/2匙油(雙壁鍋、休閒鍋或材質好的不沾鍋2匙油)，油熱後放入蒜末炒香，續入洋蔥末、紅辣椒末及3匙香菜末炒約2分鐘，倒入切碎的蝦米炒約3分鐘，再加入絞肉炒約10分鐘(材質好的鍋6分鐘)。調味料拌勻，放入拌炒均勻。
2. 河粉放入沸水中煮熟，取出放盤中，鮮蝦煮熟取出去殼。
3. 將做法1.放在河粉上，鋪上鮮蝦、撒上芹菜末、紅蔥頭，盤邊放上鳳梨，並附上碎花生、辣椒粉、白砂糖和檸檬角。

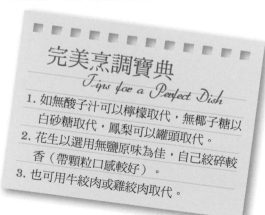

完美烹調寶典
Tips for a Perfect Dish
1. 如無酸子汁可以檸檬取代，無椰子糖以白砂糖取代，鳳梨可以罐頭取代。
2. 花生以選用無鹽原味為佳，自己絞碎較香(帶顆粒口感較好)。
3. 也可用牛絞肉或雞絞肉取代。

Ingredients

1/2 pack Thai Pho, 300g ground pork, 3T dried miniamature shrimp, 1/4 pineapple (finely-chopped), 2T minced garlic, 1/2C minced garlic, 2T minced red chili pepper, 6T minced cilantro, 3T minced celery, 3 shallots (sliced), 1/4C chopped peanut, Thai chili powder as desired, 200g fresh shrimp, white granulated sugar as needed, 1 lemon (cut into 8 equal wedges)

Seasonings

2T fish sauce, 1 1/2 lemon juice, 1T tamarind ,1T coconut sugar

Methods

1. Heat pan on medium, add 3 1/2T of cooking oil (2T of cooking for a Durotherm or good quality trying pan) and heat until smoking, stir-fry minced garlic until fragrant, add minced onion, red chili pepper and 3T of minced cilantro, then saute for 2 minutes. Add chopped dried shrimp and stir for about 3 minutes, then add ground pork about 10 minutes (6 minutes for a good quality pan). Combine the seasonings and mix well with the pork.
2. Cook noodles in boiling water until done, remove to a serving plate. Cook fresh shrimp until done, remove and discard the shells.
3. Top method (1) over the pho, spread evenly with fresh shrimp, sprinkle with minced celery and shallot, then garnish with pineapple on the side. Serve with chopped peanut, chili powder, white granulated sugar and lemon wedges as a dip.

Tips for a Perfect Dish

1. If tamarind juice is not available, use lemon. Use white granulated sugar if coconut sugar is not available. Canned pineapple may be used if fresh is not availabe.
2. Use unsalted plain peanuts in this dish, crush it yourself (A coarse texture for the ground peanut is better).
3. Ground beef or ground chicken may be used instead of pork.

材料

煎熟的鹹鮭魚
150g.、蝦米2匙、
米2杯、高湯2杯、
紅蔥頭末1匙、蒜末
1匙、南薑末1匙、
香茅2支(切段)、薑
黃粉1/2匙、紅辣椒
末1匙、香菜末1匙

做法

1. 煎好的鮭魚去皮去骨搗碎，蝦米洗淨泡水約10分鐘後切碎，米洗淨瀝乾水份。

2. 中火熱鍋，倒入1 1/2匙油，油熱後放入紅蔥頭末、蒜末爆香，加入南薑末炒香，放入蝦米炒至蝦米香味溢出，倒入米以小火炒約2分鐘，放入薑黃粉炒約2分鐘，注入高湯、放上香茅，蓋上鍋蓋轉中火煮至冒煙，改小火煮30分鐘，或將材料炒過放入電鍋內，注入高湯及香茅煮至熟，但以鍋煮較香。

3. 使用雙壁鍋或休閒鍋：中火熱鍋，倒入1 1/2匙油，油熱後放入紅蔥頭末、蒜末爆香，加入南薑末炒香，放入蝦米炒至蝦米香味溢出，倒入米以小火炒約2分鐘，放入薑黃粉炒約2分鐘，注入高湯、放上香茅，蓋上鍋蓋轉中火煮至冒煙，改小火煮6分鐘，移外鍋燜6分鐘，米即熟，香茅取出。

4. 將煮好的飯打開鍋蓋，鋪上鮭魚末，撒上紅辣椒末及香菜末。

新馬風味鹹鮭魚飯
Salty Salmon Rice

Ingredients

150g fried salty salmon, 2T dried miniature shrimp, 2C rice, 2C soup broth, 1T minced shallot, 1T minced garlic, 1T minced turmeric, 2 lemon grass leaves (cut into sections), 1/2T turmeric powder, 1T minced red chili pepper, 1T minced cilantro

Methods

1. Remove and discard bones from salmon and mince. Rinse dried shrimp and soak into water for 10 minutes, then chop finely. Rinse rice well and drain.
2. Heat pan on medium, add 1 1/2T of oil and heat until smoking, stir-fry minced shallot and minced garlic until fragrant. Add minced turmeric until flavor released, add dried shrimp and stir until flavor is released, pour in rice and stir over low heat for 2 minutes. Season with turmeric powder for 2 minutes and pour in soup broth along with lemon grass is added. Cover and increase to medium low until steaming, then reduce heat to low and cook for 30 minutes. Or put all ingredients in rice cooker and fill with soup broth along with lemon grass, cook until done. However, the dish is more tasty cooked in a pan.
3. In a Durotherm , or hot pan: heat pan on medium, add 1 1/2T of oil and heat until smoking, stir-fry minced shallot and minced garlic until fragrant. Add minced turmeric for a minute, then add dried shrimp until flavor released. Pour in rice and cook for 2 minutes, season with turmeric powder to taste. Stir for 2 minutes and pour in soup broth and add lemon grass. Cover and increase heat to medium low, cook until steaming, reduce heat to low for 8 minutes. Remove and simmer for 6 more minutes until rice is done. Discard lemon grass.
4. Remove the top and spread minced salmon on top, then sprinkle with minced red chili pepper and cilantro. Serve.

材料

米2杯、高湯2杯、去骨雞腿肉2支、香菜2支、薑片、蒜末各1匙、香茅2支

醬料

醬油膏3匙、辣豆瓣醬1/2匙、辣油1/2匙、香菜末1匙、薑末1/2匙、檸檬汁適量

做法

1. 鍋中放入1匙油，加入蒜末炒香，倒入米炒至呈象牙白色，將炒好的米和高湯放入電鍋內，鋪上香茅、雞腿、香菜頭和薑，煮約35分鐘即成。
2. 使用雙壁鍋或休閒鍋：鍋中放入1匙油，加入蒜末炒香，倒入米炒至呈象牙白色，注入高湯，鋪上香茅再放上雞腿肉、香菜頭和薑。煮至冒煙轉小火續煮6分鐘熄火，移入外鍋燜6分鐘即成。
3. 盛盤時將切好的雞肉放在盤中，醬汁拌勻淋上即可食用。

海南風味雞汁飯
Hainan Chicken Rice

Ingredients

2C rice, 2C soup broth, 2 boned chicken leg, 2 stalks cilantro, 1T ginger slices, 1T minced garlic, 2 lemon grass leaves

Sauce

3 thick soy sauce, 1/2T chili bean paste, 1/2T chili oil, 1T minced cilantro, 1/2T minced ginger, lemon juice as needed

Methods

1. Heat 1T of oil in pan, stir-fry minced garlic until fragrant, add rice and stir until the rice becomes ivory white, remove the rice along with soup broth to the rice cooker. Spread lemon grass leaves, chicken legs, cilantro roots and ginger evenly across the rice. Cook for about 35 minutes until done.
2. Use a Durotherm or hot pan: Heat 1T oil in pan, stir-fry minced garlic until fragrant, add rice and stir until the rice becomes ivory white, pour in soup broth and spread with lemon grass, chicken legs, cilantro roots and ginger. Cook until steaming, reduce heat to low and cook for another 6 minutes. Remove and simmer for 6 minutes until done.
3. Transfer to a serving plate with the chicken leg and drizzle with sauce. Mix well and serve.

CC私房美味粥之一，無法想像的好味道，可喚醒你沈睡已久的味蕾。

健康、簡易、美味，可提升免疫力，感冒時喝了這道粥很快就好了喔，

而且順口好吃沒有很重的薑味，CC的學生朋友廣為推薦，簡單到隨意煮都好吃。

One of CC's delicious private porridges. You cannot imagine how good it is. It awakens your long-sleeping tongue and stomach. This healthy, simple, and delicious meal enhances your immune system. Have this porridge dish when you are having a cold and you will soon get better. It is smooth and delicious without too much ginger flavor. CC's students highly recommend it -- it is good no matter how you make it.

舒暢京都雞肉粥
Kyoto Chicken Porridge

舒暢京都雞肉粥

材料

土雞1隻約3斤、米3杯、老薑200g.（切片拍扁）、蔥200g.、黃耆12片、白布袋1個、枸杞1大匙

塞入雞的材料：蔥3支、老薑3片、黃耆5片

調味料

鹽、白胡椒粉適量

做法

1. 將食材塞入雞內。
2. 將12枝黃耆包入白布袋中。枸杞浸泡於熱水中備用。
3. 鍋內放入雞，倒入米，加入蔥支、老薑片，和包入黃耆的白布袋，以小火煮約1小時半。
4. 使用壓力鍋：鍋中注入3公升水，蓋上壓力鍋，鍋蓋以中火煮至二條紅線，改小火續煮18分鐘即可放入調味料拌勻。
5. 撒上枸杞即可享用。

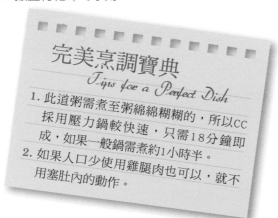

完美烹調寶典
Tips for a Perfect Dish

1. 此道粥需煮至粥綿綿糊糊的，所以CC採用壓力鍋較快速，只需18分鐘即成，如果一般鍋需煮約1小時半。
2. 如果人口少使用雞腿肉也可以，就不用塞肚內的動作。

Ingredients

1 free range chicken (approximately 3 kilos), 200g old ginger (pat flat and sliced), 200g scallions, 12 slices Radix Astragali, 1 white cotton pouch, 1T lycium berries

Stuffing for the Chicken

3 scallions, 3 slices old ginger, 5 slices Radix Astragali

Seasonings

salt and white pepper as needed

Methods

1. Stuff the ingredients inside the chicken.
2. Put 12 slices of Radix Astragali inside the pouch. Soak lycium berries in hot water for later use.
3. Place chicken in the pan along with rice, scallions, old ginger slices and the pouch with Radix Astragali. Cook over low heat for about 1 hour and 30 minutes.
4. Use a Duromatic: Fill pan with 3 liter of water, cover and cook over low until the two lines go up, reduce heat to low and cook for 18 minutes longer. Add seasonings to taste.
5. Sprinkle with lycium berries on top. Serve.

Tips for a Perfect Dish

1. This porridge dish has to be cooked until it is smooth and fine, so CC uses a Duromatic to speed the cooking time. It takes just 18 minutes to prepare. Normally it takes about 1 hour and 30 minutes with an ordinary pan.
2. If there are not enough people, just use chicken leg meat instead, then the stuffing step can be skipped.

我的學生非常喜歡我教的煎餅，配方、口感扎實而濃郁，
套一句他們的說法：老師，這真是太美味了。

My students love the pancakes that I make. The ingredients and texture
are firm and thick. As they said, "teacher, this is so good!"

廣島燒煎餅
Hiroshima Fried Cake

廣島燒煎餅

材料

蝦仁150g.(去腸泥)、透抽或花枝(切大丁狀)150g.、牛小排肉片100g.、高麗菜(切大丁狀)150g.、雞蛋麵(或油麵)200g.、美乃滋1條(100g.)、淡色醬油1/2匙、柴魚片、海苔末適量、蔥花1匙、白胡椒粉少許

麵漿

低筋麵粉2杯(過篩)、蛋3個、山藥泥1/2杯、高湯1杯、柴魚精1匙、蔥花3匙

燒烤醬

濃醬油(顏色深)1杯、味醂1/3杯、白砂糖1/4杯、小火煮約8分鐘、待涼後會呈稠狀

做法

1. 將麵漿材料全部攪拌均勻、與高麗菜、透抽(或花枝)和蝦仁一起拌勻成海鮮麵漿。將燒烤醬的材料拌勻備用。

2. 鍋中倒入1匙油、開小火、將牛小排煎至兩面呈微金黃色、取出備用。原鍋中放入蔥花炒香、加入麵條拌炒幾下、倒入白胡椒粉、醬油略拌炒、取出備用。

3. 鍋中倒入2匙油、開中火、放入海鮮麵漿煎至呈金黃色、放上煎好的牛小排、蓋上麵條；整個翻面成麵和牛小排在下方、餅皮在上方、刷上燒烤醬、擠上美乃滋、撒上海苔粉和柴魚片即成。

完美烹調寶典
Tips for a Perfect Dish

1. 燒烤醬也可以當做照燒醬、做成照燒雞腿或照燒豬排。麵糊可以當作章魚燒的麵糊。

2. 使用較大的平鍋煎會較好操作、一般家庭若沒有這麼大的平鍋、可使用長圓形的不沾鍋、或先將麵炒一下、再取出、將鍋洗好擦乾淨再煎餅。

Ingredients

150g shelled shrimp, 150g squid or cuttlefish (diced in large pieces), 100g beef short ribs, 150g cabbage (cut into large pieces), 200g egg noodles (or oiled noodles), 1 pack mayonnaise (100g),1/2T light soy sauce, bonito flakes and nori flakes as needed, 1T chopped scallion, white pepper as needed

Batter

2C cake flour (sifted), 3 eggs, 1/2C

Barbecue Sauce

1C thick soy sauce (dark), 1/3C mirin, 1/4C white granulated sugar

Methods

1. Combine the ingredients to make the batter. Add cabbage, squid and shrimp, mix until it becomes a seafood batter. Combine the ingredients of the barbecue sauce well.

2. Heat 1T of cooking oil in pan, reduce heat to low, fry beef short ribs until slightly golden and remove. Use the remaining oil to stir-fry chopped scallion until fragrant, add egg noodle and mix well. Season with white pepper and soy sauce to taste, then remove.

3. Heat 2T of cooking oil in pan, reduce heat to medium, pour in seafood batter and fry until golden. Return beef and cover with noodles. and turn over so the noodles and beef are at the bottom and the cake on top. Brush with barbecue sauce and squeeze with mayonnaise, then sprinkle with nori powder and bonito flake. Ready to serve.

Tips for a Perfect Dish

1. Barbecue sauce can be used as sauce for the Japanese teriyaki sauce used in chicken or pork chops. The batter can be used as batter for squid balls.

2. It is easier to use a bigger frying pan to fry the seafood cake. If such a big pan is not available, use an oval shaped frying pan, or stir-fry the noodles first and remove, then rinse the pan before frying the seafood cake.

日式風味咖哩雞飯
Japanese Flavored Curry Chicken Rice

這一道外面餐廳較少出現、風味不凡、簡單美味的飯料理，CC老師教完這道料理，同學回家製作的機率達九成以上，成功率也達100%，所以是相當受到歡迎的人氣食譜喔！

This dish is seldom seen in restaurants. It is an extraodinary simple but delicious rice dish. After CC taught her students, nine out of ten students made it at home, and the sucess rate is 100%. It is a very popular recipe!

日式風味咖哩雞飯

材料

雞腿肉2隻(約600g.)、米2杯、洋蔥末1/2杯、蕃茄粒罐頭1/2杯、咖哩粉3匙、奶油2匙、柳松菇罐頭1罐(去掉水份)、高湯11/2杯、巴西里少許

醃料

咖哩粉11/2匙、薑末1匙、鹽2/3匙、蒜末1匙

做法

1. 雞腿肉每隻剁約6～8塊，入醃料拌勻，放置約1小時，接著放入油鍋中，以中火煎至呈金黃色取出。
2. 奶油放入鍋內以極小火煮至融化，加入洋蔥末以小火炒透，隨即倒入咖哩粉炒至香味溢出，加入米炒約2分鐘，再放入柳松菇和做法1.炒勻，接著加入蕃茄粒和高湯拌勻，持續煮需30分鐘。
3. 雙壁鍋或休閒鍋：蓋上鍋蓋，轉中小火煮至冒煙，轉小火續煮約7分鐘，移入外鍋燜6分鐘，打開鍋蓋，撒上巴西里即成。

完美烹調寶典
Tips for a Perfect Dish

1. 咖哩要香濃好吃，一定要選用三種不同的牌子混合才會更香，因為每一個牌子的咖哩粉配方各有不同，各自選用不同的香料混合而成，而非只使用單一香料；所以多選幾種牌子混合，自然可以吃到口味豐富的咖哩。
2. 雞肉可變換為豬肉、牛肉。

Ingredients

2 chicke leg meat (approximately 600g), 2C rice, 1/2C minced onion, 1/2 C whole tomato (crushed), 3T curry powder, 2T butter, 1 can chestnut mushrooms (water discarded), 1 1/2C soup broth, parsley as needed

Marinade

1 1/2T curry powder, 1T minced ginger, 2/3T salt, 1T minced garlic

Methods

1. Chop each chicken leg meat into about 6 ~ 8 pieces, coat the marinade over on the surface evenly, let sit for one hour, then deep-fry in oiled pan over medium until golden.
2. Heat butter over the lowest flame until melt, saute minced onion on low until transparent, add curry powder and stir until the flavor released, add rice and stir for about 2 minutes. Add mushrooms and method (1) to mix and add whole tomatoes and soup broth to mix. Continue cooking for 30 minutes.
3. Use a Durotherm or hot pan: Cover and increase the heat to medium low and cook until steaming. Reduce heat to low and cook for 7 minutes more. Remove and let simmer for 6 minutes. Remove the top and sprinkle with parsley. Serve.

Tips for a Perfect Dish

1. This curry dish has to be thick and delicious. To achieve this, select three different brands and mix them together because each brand has different ingredients. Combining many brands together helps the flavor be richer and better.
2. Pork or beef may be used in place of chicken.

泰式烤雞腿蓋飯
Thai Style Roasted Chicken Over Rice

大人小孩都喜歡的萬人迷料理，
做一道即可打發一餐，是聰明主婦的最愛！

It is a dish that is loved by both adults and children. One dish
can kill one meal – it is the smart housewife's favorite.

泰式烤雞腿蓋飯

材料
去骨雞腿4隻、米2杯、高湯2杯、香茅1支(切段)、香菜2匙、蒜末2匙、檸檬葉6片、泰國甜雞醬4小碟

醃料
魚露1 1/2匙、醬油3 1/2匙(不要選太鹹的)、白砂糖2/3匙、白胡椒粉1小匙、咖哩粉1匙、椰漿1/4杯、南薑末1匙、蒜末2匙、香菜末1 1/2匙

做法
1. 醃料拌勻,放入雞腿醃一晚,使充分入味。
2. 將雞腿放入烤箱,以上下火230℃烤約15分鐘至熟(使用雙壁鍋或休閒鍋,放入2匙油,中火熱油後放入雞肉,將雞皮朝向鍋底,以中小火煎至約2分鐘半翻面,蓋上鍋蓋,轉小火續煎約6分鐘半～7分鐘)。
3. 鍋中倒入2匙油,開小火,放入蒜末炒香、隨即放入米炒約1分鐘,加入高湯、香茅和撕開的檸檬葉,轉中小火,蓋上鍋蓋待冒煙後轉小火約30分鐘(使用雙壁鍋或休閒鍋,待冒煙後轉小火煮約6分鐘後熄火,移外鍋燜6分鐘即成)。
4. 將煮好的飯盛入碗內,放上烤好的雞腿、撒些香菜。上桌時附上1小碟泰國甜雞醬。

完美烹調寶典
Tips for a Perfect Dish
1. 醃雞腿若趕時間可將醃料材料加重1/2倍,醃漬2小時。
2. 泰國甜雞醬在大型百貨公司的超市或大型量販店有售(SoGo、101、微風、家樂福)。

Ingredients
4 boneless chicken legs, 2C rice, 2C soup broth, 1 lemon grass leaf (cut into sections), 2T cilantro, 2T minced garlic, 6 lime leaf, 4 small plates Thai sweet chicken paste

Marinade
1 1/2T fish sauce, 3 1/2T soy sauce (light flavor as possible), 2/3T white granulated sugar, 1t white pepper, 1T curry powder, 1/4C coconut milk, 1T minced turmeric, 2T minced garlic, 1 1/2T minced cilantro

Methods
1. Combine the ingredients for marinade well and marinate chicken legs overnight to let the flavor be properly absorbed.
2. Roast chicken legs in oven at 230℃ on upper and lower element for about 15 minutes until done (use a Durotherm or hot pan: Heat 2T of oil in pan over medium, fry chicken legs with the skin facing down over low heat for about 2 minutes and turn over. Cover and reduce heat to low, continue frying for 6 more minutes to 7 minutes until done).
3. Heat 2T of cooking oil in pan, reduce heat to low and stir-fry minced garlic until fragrant, add rice and stir for 1 minute, pour in soup broth, lemon grass and pieces of lime leaves. Increase heat to medium low and cover. Cook until steaming and reduce heat to low for about 30 minutes. (use a double-wall pan or hot pan: Reduce heat to low when steaming, cook for 6 minutes and remove from heat. Simmer for 6 minutes until done).
4. Place the rice in bowl and top with roasted chicken leg, sprinkle with some cilantro. Serve with a small plate of Thai sweetened chicken paste on the side.

Tips for a Perfect Dish
1. If you are in a hurry, double the marinade and let sit for 2 hours.
2. Thai sweetened chicken paste can be purchased at supermarkets in the large scale department stores (such as, Sogo, Taipei 101, Breeze Center, Carrefor).

這道美味料理非常討喜，
是過年過節時餐桌上不可缺少的一道美食，
不僅可宴客、可變化成自助式的宴客方式，
也適合做成郊遊的美味便當喔！

This delicious dish pleases everybody. It is dish that one needs at the New Year's table. It can be used as a banquet dish, buffet-style party dish, or picnic lunch box.

黃金鮮蝦飯
Golden Shrimp Rice

黃金鮮蝦飯

材料

鮮蝦300g.、米2杯、高湯2杯、蒜末1匙、洋蔥末2匙、鬱金香粉1/2匙、巴西里少許

做法

1. 蝦去頭及殼留尾巴,由背部劃一下去腸泥。
2. 中火熱鍋,倒入1 1/2～2匙橄欖油,油熱後(油紋產生)放入蝦煎熟取出。
3. 原鍋放入蒜末、洋蔥末炒香,倒入洗淨的米拌炒約1分鐘,加入鬱金香粉轉中小火炒勻,倒入高湯蓋上鍋蓋,煮約30分鐘或移入電鍋(雙壁鍋倒入高湯,蓋上鍋蓋至冒煙,轉小火續煮約6分鐘,移入外鍋燜6分鐘即成)。
4. 將鍋蓋打開鋪上煎熟的蝦子、撒上巴西里即可享用。

完美烹調寶典
Tips for a Perfect Dish

米一定要炒至呈象牙白,再加入其他材料,這樣的動作可使米飯更香醇。

同場加映

宴客篇 飯做成飯丸狀擺上一隻蝦。

野餐篇 飯做成飯丸狀擺上一隻蝦後。便當盒裡放上萵苣生菜,再放上飯丸和蝦。

Ingredients

300g fresh shrimp, 2C rice, 2C soup broth, 1T minced garlic, 2T minced onion, 1/2T turmeric powder, parsley as needed

Methods

1. Remove heads from shrimp and retain the tails, devein and rinse well.
2. Heat pan over medium, add 1 1/2~2T of cooking oil until smoking, fry shrimp until done and remove.
3. Use the remaining oil to stir-fry minced garlic and onion until fragrant, then add rinsed rice and stir for about 1 minute. Add turmeric powder and reduce heat medium low, cook until evenly done and pour in soup broth. Cover and continue cooking for about 30 minutes, then remove the rice cooker (in a double-wall pan: Pour in soup broth and cover until steam comes out, reduce heat to low and continue cooking for 6 minutes. Remove and simmer for 6 minutes).
4. Remove the top and spread evenly with shrimp, sprinkle with parsley and serve.

Tips for a Perfect Dish

· The rice has to be cooked until ivory white, then add the rest of the ingredients. This extra step helps make the rice more tasty.

法式的風味征服了
嘴刁及胃口不佳的小朋友，
做法簡單卻有迷人的口感
及絕佳的奶香玉米味。

French style cooking conquers
picky people or kids who have no
appetite. The methods are simple,
yet the buttery corn aroma of the
result is fascinating.

法式奶油玉米飯
French Style Buttery Rice with Corn

法式奶油玉米飯

材 料

玉米3支、米2杯、高湯2杯、蒜末1匙、奶油11/2匙、鹽少許、巴西里少許

做 法

1. 米洗淨瀝乾水份。玉米用刨刀削下玉米粒備用。

2. 鍋中放入奶油，以中小火煮至奶油融化，放入蒜末爆香，加入米炒約2分鐘（此動作可增強米粒的香味），倒入玉米炒約1分鐘，注入高湯和少許鹽，蓋上鍋蓋煮至冒煙，轉小火續煮約30分鐘或移入電鍋（雙壁鍋或休閒鍋煮至冒煙後，轉小火續煮約時6分鐘熄火，移入外鍋燜6分鐘即成），盛盤時撒上巴西里。

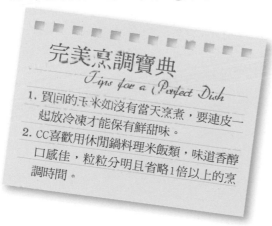

完美烹調寶典
Tips for a Perfect Dish

1. 買回的玉米如沒有當天烹煮，要連皮一起放冷凍才能保有鮮甜味。

2. CC喜歡用休閒鍋料理米飯類，味道香醇口感佳，粒粒分明且省略1倍以上的烹調時間。

同場加映

培根切約1公分寬加入，煎至呈金黃色，再與飯一起煮熟，起鍋時撒上蔥花，則又是另一道更豐富美味的飯料理。

Ingredients

3 ears corn, 2C rice, 2C soup broth, 1T minced garlic, 1 1/2T butter, a pinch of salt, parsley as needed

Methods

1. Rinse rice well and drain. Remove kernels from corn with a corn cutter.

2. Melt butter in pan over medium low, stir-fry minced garlic until fragrant, then add rice and stir for 2 minutes (this step increases the fragrance of the rice), pour in corn kernels and cook for about 1 minute. Pour in soup broth and season with salt. Cover and cook until steam comes out, reduce heat to low and continue cooking for about 30 minutes or remove to the rice cooker (in a Durotherm or hot pan: cook until steaming, reduce heat to low and continue cooking for about 6 minutes. Remove and simmer for 6 minutes until done). Transfer to a serving plate and sprinkle with parsley. Serve.

Tips for a Perfect Dish

1. Keep the corn in the refrigerator with husks on to retain its sweetness and freshness.

2. CC likes to use hot pan to prepare the rice dish. The texture is chewy and tasty and it saves half the cooking time.

Buy One Get One Free

1. Cut bacon into 1cm wide pieces and fry until golden, then cook with rice. Sprinkle with chopped scallion after removing from pan. This is another delicious dish.

這是一道外面餐廳很難吃得到的
中南美風味料理，簡單好做，好吃！

**This is a Central American dish. It is
difficult to find in a restaurant. Simple to
prepare, it is delicious too!**

波多黎各
風味燉飯
Puerto Rico Stewed Rice

波多黎各風味燉飯

材料

紅腰豆1罐、培根6片、蒜末11/2匙、洋蔥末4匙、蕃茄粒罐頭1罐、米2杯、高湯11/2杯、鹽適量、巴西里少許、奧勒岡1/2小匙

做法

1. 培根切約0.8公分寬,蕃茄粒搗碎、紅腰豆瀝掉水份。
2. 中火熱鍋,鍋中放入11/2匙油,放入培根炒至焦黃,加入蒜末、洋蔥末炒香,放入米轉小火炒約2分鐘,倒入蕃茄粒、紅腰豆、奧勒岡、高湯和鹽拌勻,蓋上鍋蓋轉中火煮至冒煙,改小火續煮6分鐘,熄火移外鍋燜6分鐘(一般鍋約30分鐘),打開鍋蓋撒上巴西里即成。

Ingredients

1 can red kidney beans, 1 1/2t minced garlic, 4T minced onion, 1 can whole tomatoes, 2C rice, 1 1/2C or T soup broth, salt as needed, parsley as needed 1/2t oregano

Methods

1. Cut bacon into 0.8cm wide pieces. Crush tomatoes. Drain red kidney beans.
2. Heat pan over medium, add 1/2T of oil and heat until smoking, fry bacon until brown, add minced garlic and onion and fry until fragrant. Reduce heat to low and saute for about 2 minutes, pour in tomatoes, red kidney beans, oregano, soup broth and salt to mix. Cover and increase heat to medium low and cook until steam comes out, reduce heat to low and cook for 6 more minutes. Remove from heat and simmer for 6 minutes (ordinary pan takes about 30 minutes), remove the top and sprinkle with parsley. Serve.

自己種香草

　　迷迭香、百里香、薄荷、羅勒、月桂葉、巴西里、鼠尾草等香草,花市都有販售。可買一小株回家自己栽種。但購買時還需買一只較大的花盆和有機土壤,將泥土倒入花盆中,挖成凹狀,把香草整株連土移入,再覆蓋上泥土。

　　半向陽性的環境比較適合栽種,水份不宜太多也不可太少。其中以薄荷最容易種,其次為迷迭香和羅勒。百里香和鼠尾草比較難種。摘取的時候要用剪刀,枝葉會繼續生長。

　　大型百貨公司超市都有賣新鮮香草,如101、SOGO和微風百貨。新鮮的薄荷、迷迭香和百里香也很適合泡香草茶來喝。

　　若買不到新鮮香草,使用乾燥的香草也可,但份量需減少1～2倍。

Plant Your Own Herbs

You can find herbs, such as rosemary, thyme, mint, basil, bay leaves, parsley, or sage in the flower market. You can buy a little stalk and plant it at home yourself. You also need to purchase a bigger flower pot and organic soil. Pour the soil into the pot, dig a hole in center and remove the whole plant including the soil, then cover with organic soil.

A half sunshine envirnment is more suitable for planting herbs. Do not water too much, or too little. Mint is the easiest of all, followed by rosemary and basil. Thyme and sage are more difficult. Remove the leaves by using a pair scissors, and the leaves will keep on growing.

Fresh herbs can easy be found in the supermarket in large department stores, such as Taipei 101, Sogo and Breeze Center. Fresh mint, rosemary and thyme are perfect for preparing herbal tea.

If fresh herbs are not available, use dried herbs. However, portion must be reduced by half or more.

Risotto是北義料理最具代表性美食，對CC而言它是引領你味蕾的好東西。

舌尖不停的跳躍著告訴你：「吃吧！吃吧！否則你會後悔！」的美味料理。

Risotto is north Italy's most representative cuisine. It is good stuff to guide your taste buds. The tip of your tongue keeps dancing around, trying to tell you, eat, eat, or you will regret it!

義大利燉飯
Stewed Risotto

義大利燉飯

材料

義大利米2杯、高湯4杯(一般鍋5杯)、培根絲100g.、帕瑪森起司粉50g.、奶油1匙、蒜末2 1/2匙、洋蔥末2匙、蘆筍100g.、南瓜丁100g.、白酒100c.c.、鹽適量

做法

1. 中火熱鍋，鍋中放入2匙橄欖油，倒入培根絲炒至呈焦黃色，加入2匙蒜末炒香，放入米以木杓炒至米呈象牙白，淋上白酒煮至酒蒸發至一半，倒入高湯至蓋過米的量，以中小火慢慢燉煮、邊煮邊攪拌，待汁收了些，再慢慢加高湯燉煮至米熟為止。
2. 另取一鍋，放入1/2匙橄欖油，加入1/2匙蒜末和洋蔥炒香，倒入南瓜丁炒至南瓜熟，以鹽調味。
3. 蘆筍剛洗淨即放入鍋內入沸水汆燙(雙璧鍋：蓋上鍋蓋以中火煮至冒煙，轉小火續煮2分鐘)，取出切丁。
4. 米煮熟後，倒入奶油和帕瑪森起司粉和南瓜、蘆筍拌勻即可盛盤。

完美烹調寶典
Tips for a Perfect Dish

1. 義大利米不可洗，否則高湯無法被米吸收。
2. 要用木杓炒拌米才不易將米破損。
3. 義大利人是吃八、九分熟的米。

Ingredients

2C risotto, 4C soup broth (measuring cup 5C), 100g shredded bacon, 50g parmesam cheese, 1T butter, 2 1/2T minced garlic, 2T minced onion, 100g asparagus, 100g diced pumpkin, 100c.c. white wine, a pinch of salt

Methods

1. Heat pan on medium, add 2T of olive oil and heat until smoking, saute shredded bacon until golden, stir in 2T of minced garlic until fragrant Pour in risotto and stir with a wooden spatula until ivory white, then drizzle with white wine and cook until the wine is half evaporated. Pour in soup broth to cover the risotto and stew over medium low, keep stirring the until soup broth is absorbed, then add more soup broth little at a time until the risotto is cooked.
2. Heat 1/2T of olive oil in another pan, saute 1/2T minced garlic and onion until fragrant, stir in diced pumpkin until done and season with salt to taste.
3. inse asparagus and blanch in boiling water (in a Durotherm: Cover and cook over medium heat until steaming, reduce heat to low and cook for 2 more minutes), remove and dice.
4. Add butter and parmesan cheese as well as diced pumpkin and asparagus to mix. Ready to serve.

Tips for a Perfect Dish
1. Do not rinse the risotto or it will not absorb the soup broth.
2. Use wooden spatula to stir to prevent destroying the rice.
3. Italians serve 80% or 90% done rice/risotto.

西班牙海鮮飯
Spainish Seafood Risotto

這道料理可是CC的招牌菜之一，CC選用義大利米來料理，

請客使出此招可說是打遍天下無敵手，

豐富的海鮮、迷人的顏色降服了全部人的胃。

This is a signature dish. CC has selected risotto to make this dish. At a banquet, this dish is a superb choice. With rich seafood ingredients and a charming color, everyone's stomach soon surrenders.

西班牙海鮮飯

材料

義大利米2杯、海鮮高湯2杯、雞高湯2杯(一般鍋再多1/2杯)、蝦8隻、淡菜8個、海瓜子200g.、墨魚150g.、扇貝6個、大蝦4隻、番紅花5g.、巴西里適量、蒜末2 1/2匙、白酒150 c.c.(一般鍋200c.c.)、培根4片切0.8公分寬、洋蔥末4匙、牛蕃茄2個(去皮切小丁)、黑胡椒、鹽、巴西里少許

做 法

1. 番紅花放入碗內,加4匙熱水浸泡10分鐘。
2. 中火熱鍋,倒入2匙橄欖油,油熱後放入培根煎至呈金黃色,加入1 1/2匙蒜末、2 1/2匙洋蔥末炒香,倒入義大利米轉小火炒至呈象牙白,改中火淋上100c.c.白酒,待酒蒸發至一半放入番紅花和牛蕃茄丁炒勻,倒入高湯至蓋住米,邊煮邊攪拌至高湯吸乾再慢慢加入高湯,邊煮邊加高湯。
3. 蝦去頭殼留尾巴,由背部劃刀,取出腸泥,大蝦由背部劃開取出腸泥。
4. 另取一鍋,放入1匙橄欖油,倒入1匙蒜末、1 1/2匙洋蔥末炒香,加入大蝦略炒幾下,再一一放入全部海鮮、撒上少許黑胡椒,淋上剩餘的白酒,略炒幾下,倒入已煮至八分熟的米再燜煮至熟,最後撒上巴西里。

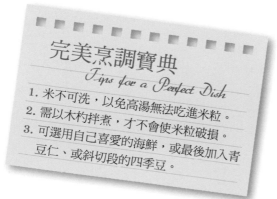

完美烹調寶典
Tips for a Perfect Dish

1. 米不可洗,以免高湯無法吃進米粒。
2. 需以木杓拌煮,才不會使米粒破損。
3. 可選用自己喜愛的海鮮,或最後加入青豆仁、或斜切段的四季豆。

Ingredients

2C risotto, 2C seafood soup broth, 2C chicken soup broth (1/2C more than usual measuring cup), 8 shrimps, 8 mussels, 200g baby clams, 150g cuttlefish, 6 scallops, 4 large shrimps, 5g saffron, parsley as needed, 2 1/2T minced garlic, 150c.c. white wine (measuring cup 200c.c.)., 4 slices bacon (cut into 0.8cm wide), 4 minced onion, 2 beefsteak tomatoes (peeled and diced finely), black pepper ,salt and parsley as needed

Methods

1. Soak saffron in hot water for 10 minutes.
2. Heat pan over medium, add 2T of olive oil and heat until smoking, fry bacon until golden, then add 1 1/2T minced garlic and 2 1/2T minced onion and fry until fragrant. Pour in risotto and reduce heat to low, stir-fry until ivory white, then increase heat to medium low and drizzle with 100c.c. of white wine. Cook until the wine is half evaporated, add saffron and beefsteak tomato to mix. Pour in soup broth to cover the risotto, cook and keep stirring at the same time until soup broth is absorbed, then add more soup broth a little at a time, cooking and adding at the same time.
3. Remove shell from shrimp and retain the tails, devein and rinse well. Devein large shrimps and rinse well.
4. Heat 1T of olive oil in another pan, stir-fry 1T minced garlic and 1/2T mince onion until fragrant. Add large shrimps to mix, then add all the seafood and sprinkle with black pepper. Drizzle the remaining white wine and saute quickly, pour in the 80% done risotto and simmer until completely done. Sprinkle with parsley and serve.

Tips for a Perfect Dish

1. Do not rinse risotto so that the soup broth can be absorbed more easily.
2. Use a wooden spatula to stir to prevent breaking apart the risotto.
3. Select seafood as desired and add peas or green beans cut diagonally into sections.

最想學會的外國菜

Part5 湯品

Soup Section

想學會的外國菜

Part5 湯品

Soup Section

最想學會的外國菜

Part5 湯品

Soup Section

To

To have a Hokk

最想學會的外國菜

Part5 湯品

Soup Section

最想學會的外國菜

Part5 湯品

Soup Section

給我一碗療癒的湯，
在夜裡、日裡，在冬天，在任何時候，
颮點秋風時來碗法式洋蔥湯吧！
輕輕淡淡去除眉間小憂愁；
寒冽嚴冬裡一定要喝花椰菜濃湯
或燉肉蔬菜湯，把胃養得暖呼呼；
春天就喝北海道農家蔬菜湯，
鄉野的風景盡入碗裡；
馬賽海鮮湯在夏日午後喝，喝完朝氣十足，
韓國宗家人參雞湯每天都來喝，
補身補心，通體舒暢！

Give me a bowl of soup for healing !!
In the night-time、day-time, in the cold winter or anytime.
To have a French onion soup in the autumn wind, you can get rid of all the sorrows.
Broccoli chowder or a beef vegetable soup in the cold winter , you can warm your stomach.
Farmer vegetable soup in the spring, you can feel the country scenery from this bowl of soup.
To have the Marseille seafood soup in the summer afternoon, you can get mighty spirit.
To have the Korean ginseng chicken soup everyday, your physical and spirit will be raised.

清爽的風味、健康的食材,這是一碗讓人心曠神怡的好湯頭。
With a clear light flavor and healthy ingredients, this is really a refreshing soup!

西班牙清湯
Spanish Clear Soup

西班牙清湯

材 料

洋蔥3個(切絲)、培根100g.(切約0.8公分小丁)、西洋芹1支(去表面粗皮切丁)、馬鈴薯2個(去皮切丁)、白酒1杯、高湯1,500c.c.(雙壁鍋1,200 c.c.)、牛蕃茄2個(切丁)、白里香1小匙、帕馬森起司2匙、巴西里適量、火腿末80g.

調味料
鹽、黑胡椒適量

做 法

1. 中火熱鍋，倒入2匙橄欖油，油熱後放入培根炒至酥脆、取出備用。

2. 利用鍋內的餘油，放入洋蔥炒透。加入西芹略炒幾下，放入馬鈴薯炒3分鐘，淋上白酒煮2分鐘。續放入百里香、高湯轉中小火，蓋上鍋蓋待冒煙後轉小火約12分鐘(雙壁鍋約煮6分鐘)。接著加入牛蕃茄及調味料調味。

3. 將湯盛入碗內，撒上培根丁、火腿末和帕馬森起司，最後撒上巴西里裝飾。

完美烹調寶典
Tips for a Perfect Dish
1. 西洋芹纖維較粗，一定要事先刮除外皮，口感才會好。
2. 馬鈴薯削皮後要泡水，才不會變黑，口感也較好。

Ingredients
3 onions (shredded), 100g bacon (cut into 0.8cm wide dices), 1 celery string (coarse surface torn and then diced), 2 potatoes (peeled and diced), 1C white wine, 1500c.c. soup broth (1200c.c. for a Durotherm), 2 beefsteak tomatoes (diced), 1t thyme, 2T parmessan cheese, parsley as needed, 80g minced ham

Seasonings
salt and pepper as needed

Methods

1. Heat pan over medium, add 2T of olive oil and heat until smoking, saute bacon until crispy and remove.

2. Use the remaining oil in pan to saute minced onion until done, add celery and stir for a minute, then add potato and cook for 3 minutes. Drizzle with white wine and cook for about 2 minutes. Continue to add thyme and soup broth, then increase heat to medium low, cover and cook until steaming. Reduce heat to low and cook for about 12 minutes (6 minutes for a Durotherm), then add beefsteak tomato and seasonings to taste.

3. Remove the soup to soup bowl and sprinkle with diced bacon, minced ham and parmesan cheese, then garnish with parlsey on the top. Serve.

Tips for a Perfect Dish
1. The celery has coarse fiber on surface. For a good texture, it must be removed.
2. Soak potato in water after peeling to prevent it from darkening. The texture will be better as well.

法式洋蔥湯
French Style Onion Soup

喝了一口會讓人溫暖舒服的湯品，尤其是浮在湯上的那一層法國麵包，
鋪滿了Guryere cheese的香味真是迷死人了！

You will just love a sip of this warm comfortable soup, especially with the
French bread floating on the top loaded with the fragrance of Gruyere cheese.

法式洋蔥湯

材料

洋蔥4個約1,000g.(切細絲)、高湯1公升、法國麵包4～8片、瑞士葛瑞爾起司100g.、奶油11/2匙、香蒜粉少許

調味料

黑胡椒、鹽適量

牛高湯

牛骨600g.、雞骨600g.、洋蔥300 g.（切大塊）、丁香4支、紅蘿蔔1/2條、西洋芹2支、月桂葉1片，新鮮百里香4支(或乾燥的1小匙)、黑胡椒粒1/2匙、大蒜6顆(帶皮)、水1,000c.c.(一般鍋1,500c.c.)一起煮沸改小火後撈出雜質漂浮沫，熬煮約2小時後(一般鍋4小時)過濾。

做 法

1. **製作牛高湯：**全部材料煮沸，瀝掉雜質和浮出的泡沫後轉小火，熬煮約4小時(雙壁鍋、休閒鍋約2小時)，再過濾。

2. 中火熱鍋，倒入11/2匙橄欖油和奶油，煮至奶油融化，放入洋蔥以小火慢慢炒透(呈茶褐色)，約60分鐘(雙壁鍋、休閒鍋40～45分鐘)。

3. 將做法2.倒入濾好的高湯中，續煮約16分鐘(雙壁鍋、休閒鍋約10分鐘)，以黑胡椒和鹽調味，盛入盤中。

4. 法國麵包撒上少許香蒜粉，放在盛好湯的湯盤上，撒上葛瑞爾起司，放入上火230℃烤至起司呈焦黃色取出，撒上巴西里。

完美烹調寶典
Tips for a Perfect Dish

法國麵包放至3天變硬後，最適合放進烤箱烤成較酥硬的口感，以上下火120℃烤約5分鐘就很好吃了。

Ingredients

4 onions(shredded) (approximately 1000g), 1 litre soup broth, 4-8 slices French bread, 100g Gruyere cheese, 1 1/2T butter, garlic powder as needed

Seasonings

black pepper and salt as needed

Beef Soup Broth

600g beef bones, 600g chicken bones, 300g onion (cut into large chunks), 4 cloves, 1/2 carrot, 2 strings celery, 1 bay leaf, 4 stalks fresh thyme (or 1t dried thyme), 1/2T black peppercorns, 6 cloves garlic (with skin on), 1000c.c. water (1500c.c. for ordinary pan)
Cook all the ingredients together until boiling, reduce heat to low and discard any impurities or bubbles on top, cook for 2 hours (4 hours for ordinary pan), then pour through a sieve.

Methods

1. To prepare soup broth: Bring all the ingredients to a boil, discard any impurities and bubbles on the surface, reduce heat to low and cook for about 4 hours (2 hours for Durotherm or hot pan), then pour through a sieve.

2. Heat pan on medium, add 1 1/2T of olive oil and butter until butter melt, saute onion over low heat for about 60 minutes until done (light brown) (40 ~ 45 minutes for a Durotherm or hot pan)

3. Pour method (2) into soup broth from method (1), continue cooking for 16 minutes (approximately 10 minutes for a double-wall pan or hot pan), season with black pepper and salt, then remove to serving bowl.

4. Sprinkle French bread with garlic powder, and put it on top of the soup bowl, sprinkle with Gruyere cheese, remove to oven and bake at 230℃ on the upper and lower element until cheese is burnt and brown. Remove and sprinkle with parsley. Serve.

Tips for a Perfect Dish

· Let French bread sit for 3 days until the texture has hardened. That is the best way to bake it in oven until crispy. Just baking at 120℃ on upper and lower element for 5 minutes will do, however.

花椰菜加上馬鈴薯，是製作濃湯的絕佳組合，

甘醇濃郁的的風味，搭配烤得香酥的法國麵包丁，真讓人垂涎三尺！

Cauliflower with potato is the best soup combination ever! Crispy diced French bread goes well with its thick, rich flavor. It really makes your mouth water!

法式花椰菜濃湯
French Style Cauliflower Cream Soup

法式花椰菜濃湯

材料

白花椰菜600g.(切小朵)、馬鈴薯600g.(去皮切丁)、洋蔥末1杯、奶水1罐、牛奶1杯、高湯800c.c.、融化奶油2匙、法國麵包1 1/2杯(切丁)、巴西里少許、黑胡椒和鹽適量

做法

1. 花椰菜洗淨放入鍋內燙熟取出(雙壁鍋或休閒鍋：蓋上鍋蓋以小火煮至冒煙隨即熄火，稍燜15秒，燜煮時不必加水，因為剛才洗淨時有水份)。
2. 奶油放入鍋內，以小火煮至融化，放入洋蔥炒透(呈淡茶褐色)，放入馬鈴薯丁煮爛後熄火待涼。
3. 將做法1.、2.一起放入調理機中，打勻成泥狀。
4. 鍋內放入奶水、牛奶、高湯，與打成泥的花椰菜糊一起煮滾，以鹽和黑胡椒調味即成。盛盤後可撒上巴西里和烤的酥脆的法國麵包丁。

完美烹調寶典
Tips for a Perfect Dish
1. 馬鈴薯去皮切丁後要先浸泡冷水，才不會變黑，且浸泡過的口味也較好吃。
2. 做法3.如不好打可加些牛奶一起打。

Ingredients

600g cauliflower (cut into small florets), 600g potato (peeled and diced), 1C minced onion, 1 can evaporated milk, 1C milk, 800c.c. soup broth, 2T melt butter, 1 1/2C French bread (diced), parsley ,black pepper and salt as needed

Methods

1. Rinse cauliflower well and blanch in pan until done, then remove (in a Durotherm or hot pan: Cover and cook over low heat until steaming, remove from heat and simmer for a minute, do not add any water during simmering because the cauliflower already has liquid after rinsing).
2. Melt butter over low heat until done, saute minced onion until done through (light brown color), add diced potato and cook until soft, then remove from heat to cool.
3. Add method (1) and (2) in a food processor and blend until mashed.
4. Heat evaporated milk, milk, soup broth in pan with mashed vegetables until boiling. Season with salt and black pepper to taste. Remove to serving plate and sprinkle with parsley and baked crispy diced French bread.

Tips for a Perfect Dish

1. Soak potato in cold water after it is peeled to prevent it from darkening. Moreover, it tastes better too.
2. In method (3) add a little milk to the food processor to make it easier to blend.

法式燉肉蔬菜湯
Fench Style Stewed Veggie Soup with Meat

豪華香濃的湯品是CC的拿手菜、也是學生們及朋友們最愛的點選料理之一，
簡單的步驟大家不妨試做看看，成功率很高喔！

**This luxurious thick soup is CC's best dish and it is also one of my students' and friends'
favorites. The cooking steps are simple. Why don't you try it – the sucess rate is pretty high!**

法式燉肉蔬菜湯

材料

牛腱600g.、雞翅腿肉300g.、白蘿蔔1小條(約200g.)、紅蘿蔔1小條(約150～200g.)、高麗菜200g.(用手撕成片狀)、四季豆100g.(斜切段)、牛蕃茄2個(切塊)、洋蔥1個(切塊)、高湯1,500cc、百里香1小匙、茵陳蒿1小匙、黑胡椒和鹽適量

做 法

1. 牛腱切約3公分厚度和雞肉均勻撒上黑胡椒、鹽(比例2：1)。
2. 中火熱鍋，倒入2匙橄欖油，油熱後放入牛腱和雞肉煎至呈金黃色取出。
3. 鍋中倒入高湯，放入洋蔥和百里香、茵陳蒿，放入煎好的牛腱燉煮約40分鐘(雙壁鍋或休閒鍋20分鐘)，放入雞肉和紅白蘿蔔續燉煮約15分鐘(雙壁鍋或休閒鍋8分鐘)，加入高麗菜煮一下，最後放入牛蕃茄和四季豆，待四季豆煮熟，以黑胡椒、鹽調味即可盛盤。

完美烹調寶典
Tips for a Perfect Dish
可以豬腱肉替代牛腱肉，但時間要稍縮短。

Ingredients

600g beef tendon, 300g chicken wing meat, 1 small daikon radish (approximately 200g), 1 small carrot (approximately 150g ~ 200g), 200g cabbage (torn into pieces), 100g green beans (cut diagonally into sections), 2 beefsteak tomatoes (cut into pieces), 1 onion (cut into pieces), 1500c.c. soup broth, 1t thyme, 1t tarragon, black pepper and salt as needed

Methods

1. Cut beef tendon into pieces 3cm thick, Sprinkle black pepper and salt evenly over beef tendon and chicken (black pepper 2:1 salt).
2. Heat pan on medium and add 2T of olive oil and heat until smoking, then fry beef and chicken until golden and remove.
3. Pour soup broth in the pan and cook onion, thyme, tarragon, and beef tendon for about 40 minutes (use a Durotherm or hot pan: 20 minutes). Add chicken, radish and carrot, continue cooking for about 15 minutes long. (8 minutes for a Durotherm and hot pan). Add cabbage and cook for a minute, add tomato and green bean to mix. Cook until green beans are done, season with black pepper salt to taste. Serve.

Tips for a Perfect Dish
· Pork tendon may be used for beef, but the cooking time is not as long.

北海道農家蔬菜湯
Hokkaido Farmer's Soup

好像走在田園的味道，清新舒暢的感受、清爽清甜原味的滋味，

讓人喝了很舒服，這是所有學生對這道湯品的感受發出的相同形容詞。

簡單豪爽的料理方式超受大家的推薦，無論如何你一定要試試看喔！

"Like walking in a field, a fresh and comfortable feeling, and a clear,
light original taste". These are the exact same compliments that the
students have about this dish. Simple, bold preparation methods are
highly recommended. You must try it!

北海道農家蔬菜湯

材 料

有機高麗菜1顆(600g.)、仿土雞腿1隻(切塊)、洋蔥3/4個(切絲)、新鮮香菇200g.、蔥花2支、高湯或水2,300 c.c.(休閒鍋1,800c.c.)

調味料

味噌3匙、鹽適量

做 法

1. 仿土雞放入滾水汆燙一下，取出瀝掉水份備用。
2. 高麗菜整棵將中間心取出，四面再各劃一刀。放入約4.5公升大的鍋內，中間放入雞，丟入洋蔥絲，注入高湯，蓋好鍋蓋煮至冒煙，轉小火煮至高麗菜軟，放入調味料拌勻，續煮約3分鐘後，加入香菇煮熟即可，起鍋前撒上蔥花。

完美烹調寶典
Tips for a Perfect Dish

1. 高麗菜用有機的較甜較可口，整棵放其甜度更佳。
2. 高湯如果用市售罐頭或雞湯粉、雞湯塊均有鹹度，所以調味時不需再放鹽。
3. 味噌不要放太多，這道料理不是味噌湯，而是加入味噌香味。
4. 味噌要加入湯內時，先放在碗內調水拌勻再倒入湯內，較易均勻不結塊。

Ingredients

1 head organic cabbage (600g), 1 free range or similar chicken leg (cut into pieces), 3/4 onion (shredded), 200g fresh shiitake mushrooms, 2 scallions, chopped, 2300c.c. soup broth or water (hot pan 1800c.c).

Seasonings

3T miso paste, salt as needed

Methods

1. Blanch chicken in boiling water and remove, rinse well and drain.
2. Remove the whole core from the center of the cabbage and score on the side four times. Remove to 4.5 litre pan along with chicken and shredded onion. Fill the pan with soup broth and cook until steaming, then reduce heat to low and cook until cabbage is soft. Season to taste, cook for another 3 minutes, then add shiitake mushrooms and cook until done. Remove and sprinkle with scallions. Serve.

Tips for a Perfect Dish

1. Organic cabbage is sweeter, putting in a whole one increases the sweetness.
2. Canned broths or chicken soup broth powders are already seasoned, so no salt is needed.
3. Do not add too much miso because this is not a miso soup dish, it is added to enhance the flavor.
4. Dissolve miso in water before adding to the pan, or it will form lumps.

馬賽海鮮湯
Marseille Seafood Soup

迷惑人心的馬賽海鮮湯，享用豐盛的海鮮料沾上馬賽醬，
還有烤得香酥的大蒜麵包，這種滋味讓人覺得人生真是美好呀！
Marseille seafood soup charms the heart. Enjoying this rich seafood cuisine dipped in
Marseille sauce with crispy garlic bread just makes you feel that life is so wonderful!

馬賽海鮮湯

材 料

白魚肉300g.(去骨)、大蝦4隻(或一般尺寸蝦12隻)、淡菜200g.、海瓜子200g.、番紅花1小撮、蕃茄粒罐頭1罐(搗碎)、法國麵包8片、大蒜奶油醬4匙、蒜末1/2匙、洋蔥末3匙、月桂葉1片、黑胡椒和鹽適量

海鮮高湯材料

大蒜6粒(切片)、洋蔥1個(切塊)、西洋芹2支(切小段)、魚骨1kg.、蝦殼1kg.、白酒1杯、水1,500c.c.(一般鍋需3,000c.c.)、月桂葉1片、黑胡椒粒1大匙、橄欖油11/2匙

馬賽醬材料

馬鈴薯2個、魚高湯5匙、紅胡椒粒1匙、蒜6粒、紅甜椒1/2個(小的1個)、橄欖油1/2匙、鹽2/3匙、巴西里1小匙

做 法

1. **自製大蒜奶油醬**：將室溫軟化的1小條奶油+1匙紅酒+1/2小匙香蒜粉+3匙蒜末+少許巴西里拌勻，即為大蒜醬，可放冰箱保存1個月(不要碰到水)。

2. **製作高湯**：中火熱鍋，倒入11/2匙橄欖油，油熱後加入大蒜和洋蔥炒透(呈淡茶褐色)，放入西洋芹略炒幾下，加入魚骨、蝦殼炒約1分鐘，淋上白酒煮至酒蒸發(1杯煮至剩1/2杯)。倒入水，加入月桂葉、黑胡椒粒熬煮約2.5小時(雙壁鍋或休閒鍋約煮1小時)，過濾後取湯汁。

3. **製作馬賽醬**：馬鈴薯洗淨不去皮，以微波爐或電鍋煮熟放入鍋內(雙壁鍋或休閒鍋：倒入1杯水(225c.c.)蓋上鍋蓋，以中火煮至冒煙，轉小火續煮約7~10分鐘，移外鍋燜10分鐘)。將馬鈴薯去皮後放入調理機中，與其他材料一起打勻成泥狀。

4. 番紅花放入碗內，倒入1/5杯溫水泡約8分鐘。

5. 鍋中倒入1匙橄欖油，開小火，油熱後加入蒜末和洋蔥末炒透，倒入海鮮高湯、搗碎的蕃茄粒、月桂葉和番紅花(連水)，煮約14分鐘(雙壁鍋或休閒鍋約煮8分鐘)。續倒入海鮮料，煮至海鮮熟，以黑胡椒和鹽調味即成。盛盤時可撒些巴西里提味，並附上馬賽醬及烤得香酥、塗上大蒜奶油醬的麵包。

Ingredients
300g white fish meat (bones removed), 4 large prawns (or 12 ordinary size shrimp), 200g mussels, 200g baby clams, 1 small pinch saffron, 1 can whole tomato (crushed), 8 slices French bread, 4T garlic flavor butter, 1/2T minced garlic, 3T minced onion, 1 bay leaf, black pepper and salt as needed

Seafood Soup Broth Ingredients
6 cloves garlic (cut into slices), 1 onion (cut into pieces), 2 strings celery (cut into small sections), 1kg. fish bone, 1kg shrimp shell, 1C white wine, 1500c.c. water (approximately 3000c.c. for ordinary pan), 1 bay leaf, 1T black peppercorns, 1 1/2T olive oil

Marseille Sauce Ingredients
2 potatoes, 5T fish soup broth, 1T red peppercorns, 6 cloves garlic, 1/2 red bell pepper (1 for small one), 1/2T olive oil, 2/3t salt, 1t parsley

Methods
1. Home-made garlic butter: Melt a small stick of butter at room temperature, then mix well with 1T of red wine, 1/2t garlic powder, 1T minced garlic and a pinch of parsley. It can be preserved in the refrigerator for 1 month (do not let it come into contact with water).

2. Soup broth: Heat pan on medium, add 1 1/2T of olive oil and heat until smoking, saute garlic and onion until done (light brown), add celery and saute for a minute, then stir in fish bone and shrimp shell. Saute for 1 minute and drizzle with white wine, cook until the wine evaporated (1C becomes 1/2C), then pour in water along with bay leaf and black peppercorns. Cook slowly for about 2 1/2 hours until done (in a Durotherm or hot pan, it takes about 1 hour). Pour through a sieve to discard any impurities.

3. Prepare Marseille Sauce: Rinse potatoes well and do not peel, cook in microwave or rice cooker until done and remove to cooking pan (in a Durotherm or hot pan: Pour 1C of water and cover, cook over medium until steam comes out, then reduce heat to low and cook for 7 to 10 minutes, remove and simmer for 10 minutes). Remove the skin from potato and blend in the food processor with the remaining ingredients until smooth and evenly blended.

4. Soak saffron in 1/5C of warm water for about 8 minutes.

5. Heat 1T of olive oil in pan over low heat until smoking, saute minced garlic and onion until done, pour in seafood soup broth, crushed tomatoes, bay leaf and saffron (along with it soaked water). Cook for about 14 minutes, (8 minutes for a Durotherm or hot pan) pour in seafood ingredients and cook until done. Season with black pepper and salt to taste. Transfer to a serving place and sprinkle with parsley to enhance the flavor. Serve with Marseille sauce and crispy garlic bread on the side.

牛肉的香醇夾帶著蔬菜的甘甜味和帶酸的濃醇蕃茄口感，
這是CC最愛的蔬菜湯，讓人有溫暖的戀愛感。

This is CC's favorite vegetable soup. The aroma
of the beef, the sweetness of the vegetables,
and the sour of the tomato will give you the
warm feeling of falling in love.

匈牙利
牛肉蔬菜湯
Hungarian Beef Veggie Soup

匈牙利牛肉蔬菜湯

材 料

牛腩(牛肋)400g.、大蒜8粒、洋蔥2個(1個切丁、1個切絲)、西洋芹2支(1支切段、1支削皮切丁)、紅蘿蔔丁1/2杯(保留削剩的皮及頭尾)、牛蕃茄丁1杯(去皮)、高麗菜丁1 1/2杯、蕃茄糊2匙、香蒜粉1/2小匙、匈牙利紅椒粉1/2匙、牛高湯(參考P.119)2,000c.c. (雙壁鍋或休閒鍋1,500 c.c.)

調味料

黑胡椒、鹽適量

做 法

1. 牛腩切塊約3.5公分長,撒上黑胡椒、鹽(比例2:1)及香蒜粉,醃置約30分鐘。
2. 中火熱鍋,倒入2匙橄欖油,油熱放入醃好的牛腩切塊煎至呈金黃色取出。
3. 鍋中續放入1 1/2匙橄欖油,放入大蒜爆香,加入洋蔥絲炒透,倒入西芹段、紅蘿蔔削剩的皮及頭尾炒約5分鐘,放入蕃茄糊炒香,倒入3杯高湯煮約30分鐘(雙壁鍋或休閒鍋15分鐘),過濾去除掉殘渣。將高湯倒入做法2.,燉煮約1.5小時(雙壁鍋或休閒鍋30分鐘)。
4. 鍋中放入1 1/2匙橄欖油(雙壁鍋或休閒鍋1匙),倒入洋蔥丁炒透後放進西芹丁、紅蘿蔔丁、高麗菜丁、牛蕃茄丁炒熟,倒入做法3.續煮10分鐘(雙壁鍋或休閒鍋5分鐘),最後放入匈牙利紅椒粉煮一下,再以黑胡椒、鹽調味。

完美烹調寶典
Tips for a Perfect Dish

1. 如不吃牛肉可改用豬腱肉。
2. 蕃茄糊開罐後沒用完,要將蕃茄糊全部取出,置放保鮮盒內或冷凍,如不取出放在罐頭內容易發黴。

Ingredients

400g beef finger (rib), 8 cloves garlic, 2 onions (1 diced, 1 shredded), 2 strings celery (1 cut into sections, 1 peeled and diced), 1/2C diced carrot (retain the peels as well head and end), 1C diced beefsteak tomato (peeled), 1 1/2C diced tomato, 2T tomato paste, 1/2t garlic powder, 1/2T Hungarian red chili powder, 2000c.c. beef soup broth (see p.119] for reference) (1500c.c. for Durotherm or hot pan)

Seasonings

black pepper and salt as needed

Methods

1. Cut beef rib fingers into pieces 3.5cm long, sprinkle with black pepper and salt (2:1) as well as garlic powder, let sit for 30 minutes until flavor is absorbed.
2. Heat pan over medium, add 2T of olive oil until smoking, fry beef until golden and remove.
3. Add 1 1/2T of olive oil in pan, saute garlic until fragrant, add onion until done through, then add celery sections, carrot peel, head and end. Stir for about 5 minutes, then pour in tomato paste and 3C of soup broth. Cook for about 30 minutes (15 minutes for a Durotherm or hot pan), remove any impurities. Combine with beef soup broth and pour into method (2). Stew for about 1.5 hours until done (30 minutes for a Durotherm or hot pan).
4. Heat 1 1/2T of olive oil and stir-fry onion until done through, add diced celery, carrot, cabbage and beefsteak tomato until cooked. Add to method (3) and cook for 10 minutes longer (5 minutes for Durotherm or hot pan). At the end, season with Hungarian red chili powder and cook for a minute, then season with black pepper and salt to taste. Ready to serve.

Tips for a Perfect Dish

1. Use pork tendon if beef is not desired.
2. If the tomato paste is not used up, pour it into a container and freeze, or it will easily spoil.

俄羅斯羅宋湯
Borscht

俄羅斯羅宋湯是非常經典的傳統料理，
搭配酸奶是最美味、最令人迷戀的口感！

Russian Borsht is a classic dish made with delicious
sour cream. You will become obsessed with it!

俄羅斯羅宋湯

材料

牛腩(牛肋)600g.、甜菜300g.、洋蔥1/2個(切塊)、馬鈴薯大1個或小2個(去皮切塊)、月桂葉1片、蒜末1匙、洋蔥1個(切絲)、紅蘿蔔1/2條(切絲,保留削剩的皮及頭尾)、高麗菜150g.(切絲)、牛蕃茄小的2個(去皮切塊),水1,500c.c.(一般鍋2,200c.c.)、黑胡椒粒1/2匙、奶油11/2匙、酸奶4匙、巴西里適量

調味料
黑胡椒、鹽適量

做 法

1. 為保有味道和防止顏色流失,甜菜整顆帶皮放入沸水煮約1小時10分鐘(雙壁鍋或休閒鍋約40分鐘)。冷卻後去皮(在碗內以免染上顏色)、搗碎,備用。

2. 牛肉切3.5公分左右放入沸水中,煮沸後瀝掉雜質和浮出的泡沫,放進洋蔥塊、紅蘿蔔削剩的皮和頭尾、月桂葉和黑胡椒粒,煮滾後轉小火續煮約60分鐘(雙壁鍋或休閒鍋約25分鐘),將牛肉取出湯汁過濾。

3. 鍋中放入奶油,以小火煮至融化,放入蒜末炒香、加入洋蔥絲炒透(呈淡茶褐色),隨即加入紅蘿蔔絲略炒一下,再放入高麗菜絲、牛蕃茄,倒入做法2.的高湯、牛肉和馬鈴薯塊。放入搗碎的甜菜,續煮約20分鐘(雙壁鍋或休閒鍋約10分鐘),以黑胡椒鹽調味。

4. 盛入碗內,上面放上1匙酸奶、撒上巴西里即可享用。

完美烹調寶典
Tips for a Perfect Dish

1. 牛肉可選用牛尾。
2. 新鮮甜菜較費時,可選用罐頭甜菜,罐頭烹煮即可省略做法1.的步驟,直接取出去水搗碎。也可直接改用蕃茄粒罐頭。

Ingredients
600g beef rib fingers, 300g beets, 1/2 onion (cut into pieces), 1 large potato or 2 small potatoes (peeled and cut into pieces), 1 bay leaf, 1T minced garlic, 1 onion (shredded), 1/2 carrot (shredded, retain skin, both head and end), 150g cabbage (shredded), 2 small beefsteak tomatoes (peeled and cut into small pieces), 1500c.c. water (2200c.c. for ordinary pan), 1/2T black peppercorns, 1 1/2T butter, black pepper and salt as needed

Methods
1. To prevent the flavor and color from being lost, cook the whole beet with the skin on in boiling water for about 1 hour and 10 minutes (40 minutes for a Duromatic or hot pan), remove and cool, then remove the skin (in a bowl to prevent the color from bleeding out, and crush.
2. Cut beef into 3.5cm long and cook in boiling water until boiling, discard any impurities and foam floating on the surface. Add onion, carrot peel, head and end, bay leaf and black pepper. Bring to a boil, then reduce heat to low and continue cooking for 60 minutes longer, (25 minutes for a Duromatic or hot pan), then remove beef and pour the soup through sieve to remove any impurites.
3. Heat butter in pan on low until melted, saute minced garlic until fragrant, add shredded onion until done (light brown), then add shredded carrot to mix. Add shredded cabbage, beefsteak tomato as well as soup broth from method (2), beef and potato to mix. Next, put in crushed beet and continue cooking for about 20 minutes longer (10 minutes for a Duromatic or hot pan), season with black pepper to taste.
4. Transfer to a soup bowl, top with 1T of sour cream and sprinkle with parsley. Serve.

Tips for a Perfect Dish
1. Beef tail may be used instead of beef fingers.
2. If fresh beet takes too much trouble, use canned beet instead. If you use canned beet, skip method (2), just remove the beet from the can and crush.If beet is not available, use a whole can of whole tomatoes.

韓國宗家
人參雞湯
Korean Ginseng Chicken Soup

不喜歡人蔘味道的人都會因此道湯的湯頭味而喜歡。

Even those who do not like the flavor of ginseng love this soup dish.

韓國宗家人參雞湯

材料
土雞1隻、新鮮人參(水參)2支、糯米2匙、紅棗10粒、大蒜12粒、乾栗子10粒、乾百果1/4杯、白布袋1個、牙籤3支、高湯或水約3,000c.c(壓力鍋2,500c.c.)

調味料
鹽、白胡椒粉各適量

人參雞湯沾醬
韓國辣椒醬2匙、醬油1匙、洋蔥泥1/2匙、蒜末1匙、香油1匙、白砂糖1小匙、蔥末1/2匙、檸檬汁1/2匙、白芝麻

做法
1. 雞洗淨拭乾水份，肚內塞入4粒紅棗、4粒大蒜、4粒栗子和8粒百果，以牙籤剌好。
2. 將糯米放入白布袋內綁好。
3. 做法1.放入鍋內，倒入高湯或水，放入糯米袋和剩餘的大蒜、栗子、百果和鮮人參，煮至沸騰，轉小火續煮約1小時20分鐘，打開鍋蓋放入紅棗再繼續煮約6分鐘，放入鹽和白胡椒粉調味即成，上桌時附上拌勻的人蔘雞湯沾醬。
4. 使用壓力鍋：放入做法1.，倒入高湯或水，放入糯米袋和剩餘的大蒜、栗子、百果，和鮮人參，蓋上壓力鍋鍋蓋，上升2條紅線轉小火煮18分鐘洩壓後，打開鍋蓋放入紅棗續煮約3分鐘，加入鹽和白胡椒粉調味，上桌時附上拌勻的人蔘雞湯沾醬。

完美烹調寶典
Tips for a Perfect Dish
1. CC家常將剩餘的湯再放入些蒜粒煮一下，再加入陽春麵煮好，撒些蔥花美味哦！
2. 白布袋可在中藥行買到。
3. 人參雞湯沾醬用來醃肉也非常好吃，也可當拌乾麵醬。

Ingredients
1 free range chicken, 2 fresh ginseng (water ginseng), 2T sticky rice, 10 red dates, 12 cloves garlic, 10 dried chestnuts, 1/4C dried gingkoes, 1 white pouch, 3 toothpicks, 3000c.c. approximately soup broth or water (2500c.c. for Duromatic)

Seasonings
salt and white pepper as needed

Dipping Sauce for Ginseng Chicken Soup
2T Korean chili paste, 1T soy sauce, 1/2T grated onion, 1T minced garlic, 1T sesame oil, 1t white granulated sugar, 1/2T minced scallion, 1/2T lemon juice, white sesame seeds as needed

Methods
1. Rinse chicken and dry well. Stuff 4 datso, 4 cloves garlic, 4 dried chestnuts and 8 gingkos inside the chicken stomach, then seal with toothpicks.
2. Put sticky rice in the white pouch and tight up well.
3. Remove method (1) to pan along with soup broth and water added. Add sticky rice pouch and remaining garlic, chestnuts, gingko and fresh ginseng.Bring to a boil, reduce heat to low and continue cooking for 1 hour and 20 minutes . Remove the top and add dates. Cook for 6 minutes long, season with salt and white pepper to taste. Serve with dipping sauce on the side.
4. Use a Duromatic: Place chicken in pressure cooker along with soup broth and water added. Add sticky rice pouch and remaining garlic, chestnuts, gingko and fresh ginseng. Cover and cook until two red lines go up, reduce heat to low and. Cook for 18 minutes, then add red dates and cook for 3 more minutes. Season with salt and white pepper to taste. Serve with dipping sauce on the side as a dip.

Tips for a Perfect Dish
1. CC's family always adds garlic cloves to the leftover soup, then adds a little white noodles and then sprinkles it with chopped scallion. This can be a delicious simple meal.
2. The white pouch can be purchased at the Chinese medicine store.
3. The dipping sauce can also be used for marinating pork. It is delicious. It can be used as a mixing sauce for dried noodles.